Signaling Pathways in Cancer Pathogenesis and Therapy

David A. Frank

Editor

Signaling Pathways in Cancer Pathogenesis and Therapy

 Springer

Editor
David A. Frank
Dana-Farber Cancer Institute
Harvard Medical School
Boston, MA, USA
david_frank@dfci.harvard.edu

ISBN 978-1-4899-9222-2   ISBN 978-1-4614-1216-8 (eBook)
DOI 10.1007/978-1-4614-1216-8
Springer New York Dordrecht Heidelberg London

Printed on acid-free paper

Springer is part of Springer Science+Business Media (www.springer.com)

Contents

Contributors

Robert T. Abraham Center for Integrative Biology and Biotherapeutics, Pfizer Biopharmaceuticals, Pearl River, NY, USA

Arijit Chakravarty Department of Oncology, Millennium Pharmaceuticals, Inc., Cambridge, MA, USA

Natalie D'Amore Department of Oncology, Millennium Pharmaceuticals, Inc., Cambridge, MA, USA

Jeffrey A. Ecsedy Department of Oncology, Millennium Pharmaceuticals, Inc., Cambridge, MA, USA

David A. Frank Department of Medical Oncology, Dana-Farber Cancer Institute, Harvard Medical School, Boston, MA, USA

Gary E. Gallick MD Anderson Cancer Center, Houston, TX, USA

Thanos D. Halazonetis Departments of Molecular Biology and Biochemistry, University of Geneva, Geneva, Switzerland

S. Percy Ivy Investigational Branch, Cancer Therapy Evaluation Program, National Cancer Institute, NIH, Rockville, MD, USA

Lawrence Lum Investigational Branch, Cancer Therapy Evaluation Program, National Cancer Institute, NIH, Rockville, MD, USA

Mark Manfredi Department of Oncology, Millennium Pharmaceuticals, Inc., Cambridge, MA, USA

Vito J. Palombella Infinity Pharmaceuticals, Inc., Cambridge, MA, USA

Margaret A. Read Infinity Pharmaceuticals, Inc., Cambridge, MA, USA

Naoko Takebe Investigational Branch, Cancer Therapy Evaluation Program, National Cancer Institute, NIH, Rockville, MD, USA

Sarah R. Walker Department of Medical Oncology, Dana-Farber Cancer Institute, Harvard Medical School, Boston, MA, USA

Jon R. Wiener MD Anderson Cancer Center, Houston, TX, USA

Chapter 1
Signaling Pathways in Cancer: Twenty-First Century Approaches to Cancer Therapy

David A. Frank

Although descriptions of the disease we now call cancer have been found in ancient writings, useful treatments for malignancies have only been available since the 1940s. The work of Goodman and Gilman at Yale on alkylating agents, and of Sidney Farber in Boston on antifolates, allowed for the first time the reliable regression of advanced cancers, largely leukemias and lymphomas. It soon became apparent that single anticancer agents would generally only lead to transient responses, and as the tumors recurred, they were resistant to repeated treatments with the same agents. Thus, the era of combination chemotherapy arrived, with carefully designed clinical trials, often spearheaded at the National Cancer Institute, testing the effects of various combinations of chemotherapy agents. It was soon apparent that by using complementary mechanisms of action, and avoiding the emergence of resistance, multiagent chemotherapy was considerably more effective than single agents, and previously lethal leukemias and lymphomas could be cured.

In subsequent decades, drugs that targeted other cellular components, such as microtubules and topoisomerases, were added to the armamentarium. In the 1970s, diseases that had been rapidly fatal, like advanced testicular cancer, were now eminently curable. Advances in supportive care, including transfusion of blood products, antibiotic support, and antinausea drug furthered our ability to treat patients with cancer. However, in the 1990s, at the time of celebrations of the 50th anniversary of some of the seminal moments in the discovery of anticancer agents, we had reached somewhat of a plateau. Relatively few new anticancer agents were emerging, and those that were being approved were often just analogues of prior agents. For some cancers, like acute myelogenous leukemia (AML), the most lethal form of leukemia in adults, we were still using the same two cytotoxic agents we had

D.A. Frank (✉)
Department of Medical Oncology, Departments of Medicine,
Dana-Farber Cancer Institute, Brigham and Women's Hospital
and Harvard Medical School, 450 Brookline Avenue, Boston, MA 02215, USA
e-mail: david_frank@dfci.harvard.edu

D.A. Frank (ed.), *Signaling Pathways in Cancer Pathogenesis and Therapy*,
DOI 10.1007/978-1-4614-1216-8_1, © Springer Science+Business Media, LLC 2012

been using for decades. Advanced forms of common cancers, such as cancers of the lung, breast, prostate, colon, and pancreas remained incurable, and approximately 500,000 people were dying of cancer per year in the United States.

However, as the twentieth century was ending and the twenty-first century was beginning, a very different approach to cancer therapy was being reported in both scientific journals and local newspapers. A new treatment had emerged for a relatively rare blood cancer, chronic myelogenous leukemia (CML). Decades of research had shown that nearly every patient with CML had a translocation between chromosomes 9 and 22, leading to the fusion of two genes, *Bcr* and *Abl*, leading to the production of a chimeric protein, Bcr/Abl. This was a highly active tyrosine kinase that phosphory-lated a range of cellular substrates, and drove the malignant behavior of the leukemic cell. Through a combination of solid scientific work, clinical acumen, and personal drive, Brian Druker and colleagues developed a drug, imatinib mesylate, often referred to by its trade name, Gleevec. Imatinib, a pill taken once daily, inhibited the tyrosine kinase activity of Bcr/Abl, and rapidly reversed the signs and symptoms of leukemia in the great majority of CML patients who took it. Although "miracle cures" seem to occur only in movies, for many patients with CML, imatinib was truly miraculous.

The success of imatinib raised great hope that other cancers would be vanquished in a similar fashion. In some rare cancers, like gastrointestinal stromal tumor (GIST), an activating mutation in another kinase, c-kit, was found, and patients with these mutations often had a dramatic response to imatinib and other tyrosine kinase inhib-itors. Subtypes of common cancers were also found to have mutations that could be exploited therapeutically, like Her2 amplification in breast cancer (which can be treated by both drugs that block its activity and antibody-based therapies) or muta-tion of the epidermal growth factor receptor (EGFR) in non-small cell lung cancer (which can be treated with drugs blocking its inappropriately activated kinase).

These triumphs represented the fruits of years of basic research focused on uncov-ering the molecular underpinnings of cancer. However, we still have 500,000 Americans dying each year of cancer, and the challenge now is to extend this para-digm of basic discovery being translated into effective therapies. It is with that back-ground in mind that this volume is particularly timely. The goal was to recruit experts on many of the key pathways whose function or diversion plays a critical role in the biology of a cancer cell, with a particular thought as to how one can then use their knowledge to consider therapeutic applications that can be offered to patients. Recognizing that every scientific sector has much to contribute in this area, a multi-national team of authors, working in industry, government, and academia was asked to highlight key areas for a twenty-first century approach to cancer therapy, based on an intimate knowledge of the workings and derangements of a cancer cell.

Each of the chapters in some ways weaves together basic biology and early approaches to cancer therapy with the most current and sophisticated approaches being developed. Starting with a focus on antimitotic agents, we start with a consider-ation of tubulin-targeting agents, such as vinca alkaloids, which represent some of the first anticancer agents given to patients, and end with drugs targeting specific kinases and other enzymes that regulate key steps in mitosis. This is followed by a review of the signaling events surrounding DNA damage which provides insight into both the pathogenesis of cancer, and unique ways in which cancer cells could be targeted.

The next chapter also takes a historical perspective, starting with observations made by Peyton Rous on animal tumors in the early 1900s to our current understanding of the role that Src and its related tyrosine kinases play in normal cellular function and tumor pathogenesis, and as targets for current cancer therapy.

Reflecting on the importance of basic biologic research, including developmental studies in "lower" organisms, we now understand that pathways named for phenotypes in Drosophila, such as Wingless and Hedgehog, are important in tissue homeostasis in mammals, and in the development of cancer in humans. Once again, this knowledge opens up a number of opportunities for targeted rational therapy for patients, which has the potential to be both more effective and less toxic.

While identification of the mutations occurring in a cancer cell will hopefully lead to therapies directly targeting these molecular events, such as imatinib for CML, most common human cancers have a large number of mutations, and it can be difficult to deconvolute which are of critical importance, and exactly how they drive malignant cellular behavior. However, these mutations often lead to the activation of signaling pathways which converge on a relatively small number of transcription factors, such as STATs. While STATs themselves are not mutated in cancer, by integrating signals from multiple pathways, they represent excellent targets for cancer therapy.

Finally, as biological research uncovers targets that might be particularly useful in treating cancer, the key question arises as to how can one take this knowledge and actually develop a therapeutic agent that can be given to a patient. The final chapter was written by Michael Corbley, a uniquely talented scientist who has comprehensively reviewed the broad topic of protein therapeutics for cancer, an exciting and dynamic area of therapeutic research. Amazingly, Michael wrote this chapter while he himself was battling advanced cancer. Tragically, Michael died shortly after completing this work. In some ways, Michael's courage, strength, and commitment encapsulates where we are with cancer therapy in the second decade of the twenty-first century. We have wonderfully talented and dedicated researchers who are putting their enormous talents to work at the interface of scientific discovery and clinical medicine. At the same time, we have incredibly strong and brave patients who very much need more effective, less toxic, rationally designed cancer therapies. Through both Michael's wisdom shared in these pages and the inspiration of his own battle with this disease, it is hoped that this volume will provide another step upward toward our shared goal of making cancer an eminently controllable disease, and thus it is to Michael Corbley that this book is dedicated.

Chapter 2
Current and Next Generation Antimitotic Therapies in Cancer

Jeffrey A. Ecsedy, Mark Manfredi, Arijit Chakravarty, and Natalie D'Amore

2.1 Current Therapeutic Application of Antimicrotubule Agents

The neatly ordered, symmetrical appearance of the microtubule spindle during mitotic cell division belies the highly dynamic nature of this critical event during mitosis. In organizing the mitotic spindle and executing a successful division, a wide array of proteins cooperate to line up and then move chromosomes along their microtubule scaffolds (Fig. 2.1). The disruption of the mitotic machinery as a chemotherapeutic approach therefore has the potential to cause cancer cell death or arrest without affecting normal, nondividing tissue. Traditional antimitotic agents comprise those that directly interfere with microtubule dynamics, essential for mitotic spindle assembly and the subsequent alignment and segregation of DNA to daughter cells. Antimicrotubule agents currently being used in clinical setting are the taxanes, vinca alkaloids, and epothilones. These agents are used in a host of cancer types as single agents and in combination with other oncology therapeutics.

Paclitaxel (brand name Taxol), the first taxane identified, was discovered in extracts of bark from the Pacific yew tree in the early 1960s and was approved for the treatment of ovarian cancer three decades later in 1992. Docetaxel (brand name Taxotere) is a semisynthetic derivative of paclitaxel that is more soluble and has demonstrated distinct clinical activity in some cancers, including metastatic breast cancer (Jones et al. 2005). In general, paclitaxel and docetaxel have a similar spectrum of clinical activity including ovarian, lung, breast, bladder, and prostate cancers. Even though both paclitaxel and docetaxel have been used clinically for many years, their utility continues to expand into new indications and in new combinations with other agents.

J.A. Ecsedy (✉) • M. Manfredi • A. Chakravarty • N. D'Amore
Department of Oncology, Millennium Pharmaceuticals, Inc.,
40 Landsdowne Street, Cambridge, MA 02139, USA
e-mail: Jeffrey.Ecsedy@mpi.com

D.A. Frank (ed.), *Signaling Pathways in Cancer Pathogenesis and Therapy*,
DOI 10.1007/978-1-4614-1216-8_2, © Springer Science+Business Media, LLC 2012

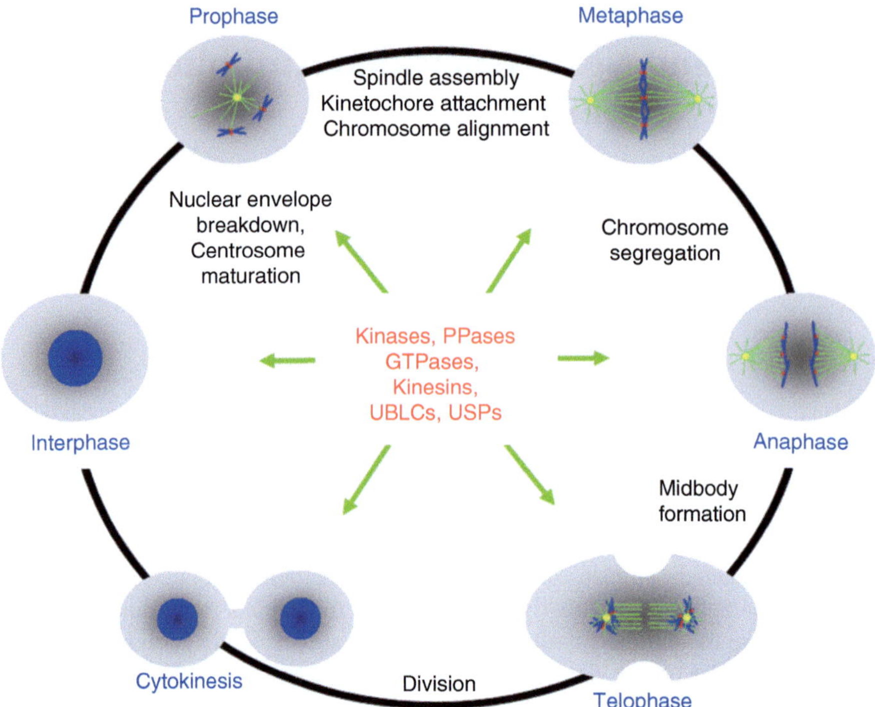

Fig. 2.1 Overview of normal progression through mitosis. A diverse array of kinases, phophatases (PPases), GTPases, kinesins, ubiquitin-like conjugators (UBLCs), and ubiquitin specific proteases (USPs) orchestrate the various stages of mitosis; including prophase, metaphase, anaphase, telophase, and cytokinesis. Some of the critical events that occur during each of these stages are highlighted

Abraxane™ is paclitaxel formulated in albumin-bound nanoparticles, eliminating the need for Cremephor-EL in the formulation, a vehicle that on its own has demonstrated toxicities and necessitates premedication (Ibrahim et al. 2002). Abraxane was approved on clinical data that demonstrated greater activity and safety than paclitaxel in patients with metastatic breast cancer.

The toxicities associated with each of the taxanes are similar, and include neutropenia as the major dose limiting toxicity, along with significant peripheral neuropathy. In fact, dose reductions are frequent in heavily pretreated patients to mitigate the severity of these toxicities. Interestingly, in clinical studies dose reductions did not reduce the clinical response of the agents, suggesting that the optimal biological dose may be lower than the maximum tolerated dose (Salminen et al. 1999). Weekly administration of the taxanes has become more frequently used as clinical data demonstrated less myelosuppression with no decrease in clinical response (Gonzalez-Angulo and Hortobagyi 2008). Interestingly, in breast cancer studies, weekly paclitaxel showed better response rates than once every 3 week dosing (Seidman et al. 2008). However, weekly paclitaxel has demonstrated greater neuropathy than the every 3 week schedule.

The vinca alkaloids where discovered in the 1950s from extracts of the leaves of the periwinkle plant (Catharanthus roseus). The vinka alkaloids were originally considered for use as antidiabetic agents, however, it was quickly learned that they possessed antiproliferative activity. Vincristine and Vinblastine, both microtubule destabilizers are the oldest and most studied members within this class of microtubule binding agents, and are now standard of care agents in various cancer types. Vincristine is used for treating several tumor types, including Non-Hodgkin and Hodgkin lymphoma and certain pediatric cancers, while vinblastine is used for treating testicular, Hodgkin lymphoma, lung, head, and neck, and breast cancer. More recently vinorelbine, a semisynthetic vinca alkaloid, was discovered to have a better preclinical profile than other family members (Krikorian and Breillout 1991). Vinorelbine was approved for treating NSCLC and has shown promising activity in breast, head and neck, ovarian, and squamous cell carcinoma (Burstein et al. 2003; Jahanzeb et al. 2002). Toxicities associated with the various vinka alkaloid members are similar, with neutropenia and peripheral neuropathy being dose limiting.

The epothilones are a newer class of tubulin binding agents that were first isolated in the 1990s from the myxobacterium *Sorangium cellulosum* (Bollag et al. 1995). There are several naturally occurring (epothilone A, B, C, and D) and semisynthetic variants currently under clinical investigation, with Ixabepilone, a derivative of epothilone B, now approved for the treatment of advanced breast cancer (Fumoleau et al. 2007). Similar to the taxanes, the epothilones promote microtubule stability, and in fact share the same binding site with paclitaxel. The perceived advantages over the taxanes include greater potency and decreased likelihood for resistance resulting from drug pumps and tubulin mutations (Kowalski et al. 1997; Wartmann and Altmann 2002). Moreover, the epothilones are formulated in vehicles that are better tolerated than the cremophor used for paclitaxel (Sessa et al. 2007; Watkins et al. 2005).

There are several differences in the toxicities and clinical activity between the various epothilones. Patupilone is the natural product epothilone B and is in phase III studies versus doxorubicin in ovarian, fallopian tube, and peritoneal cancers. Patupilone demonstrated Phase II single agent activity in several tumor types including colorectal, gastric, hepatocellular, non-small cell lung cancer, ovarian, and renal cancer (Harrison et al. 2009). Unlike the taxanes and other epothilones, diarrhea rather than neutropenia was the major dose limiting toxicity in all the schedules tested (Rubin et al. 2005). Interestingly, there was little neutropenia or significant peripheral neuropathy seen in the trials.

Ixabepilone is a derivative of epothilone B which has greater metabolic stability than the parent natural product. Ixabepilone was approved from a phase II study as a single agent for patients with advanced breast cancer who are resistant to prior treatment with an anthracycline, taxane, and capecitabine (Perez et al. 2007). Ixabepilone has demonstrated activity in bladder, breast, non-Hodgkin lymphoma, non-small cell lung cancer, pancreatic, prostate, renal, and sarcoma (summarized in (Harrison et al. 2009)). Unlike patupilone, in a phase II study ixabepilone failed to demonstrate activity in colorectal cancer suggesting that these agents may have a different spectrum of clinical activity. Ixabepilone completed a pivotal phase III trial in advanced breast cancer in combination with capecitabine where it demonstrated greater activity than capecitabine alone (Thomas et al. 2007). Particularly interesting was the improved progression free survival in the combination group in patients with triple negative breast cancer, a patient population that has a high unmet medical need. The dose limiting toxicities in the majority of the trials were neutropenia and fatigue. The epothilones represent a promising new class of tubulin-binding antimitotics that have already differentiated themselves from the taxanes.

2.2 Antimitotic Agents: Mechanism of Action

Inhibition of the mitotic machinery results in a diverse array of outcomes, primarily leading to cell death or arrest (Fig. 2.2). As the effect of antimitotic agents is not limited to cancer cells alone, the dose-limiting toxicities of these drugs in a clinical setting frequently manifest in rapidly dividing tissue and are often accompanied by severe peripheral neuropathy in the case of antimicrotubule agents. Therefore, the narrow therapeutic index of antimitotic agents necessitates a precise understanding of the mechanism of action of these drugs to maximize the chances of rational development of these therapies.

Our understanding of the basic science underlying antimitotic therapies has been primarily developed using taxanes, including paclitaxel and docetaxel. Taxanes stabilize microtubules by altering the kinetics of microtubule depolymerization. In mammalian cells grown in culture, high concentrations of paclitaxel cause the aggregation of microtubules (Schiff and Horwitz 1980). At lower concentrations that resemble exposures achieved in clinical settings, the primary effect of paclitaxel is to stabilize microtubules, and thereby dampen the dynamic instability of microtubules that is a requisite for efficient spindle assembly. As a result of this dampening, microtubules are unable to grow and shrink rapidly, and their ability to bind to condensed chromosomes during mitosis is compromised. Efficient chromosome alignment is thus affected, and this failure of chromosome alignment leads to mitotic delays mediated via the spindle assembly

Fig. 2.2 (continued) and their inhibition can lead to delayed mitotic entry. Once in mitosis, perturbation of a variety of targets leads to dramatic abnormalities in centrosome maturation/separation, mitotic spindle formation, chromosome condensation, attachment of microtubules to kinetochores, and spindle assembly checkpoint signaling among other events, leading to chromosome alignment defects. The fate of these cells is varied, and can include apoptosis directly from mitosis, anaphase initiation accompanied by chromosome segregation defects leading to an aneuploid division, or exit from mitosis without cytokinesis via mitotic slippage leading to G1 tetraploid cells (double the normal DNA content at this stage). The interphase cells derived from these abnormal mitotic divisions often present as micronucleated or multinucleated. G1 tetraploid cells may undergo additional rounds of DNA replication via a process referred to as endoreduplication resulting in polyploid cells. Ultimately, these cells will eventually die via apoptosis or become senescent, which themselves can eventually undergo apoptosis. Lastly, if cells survive the events associated with an abnormal division, they can undergo additional rounds of mitotic division

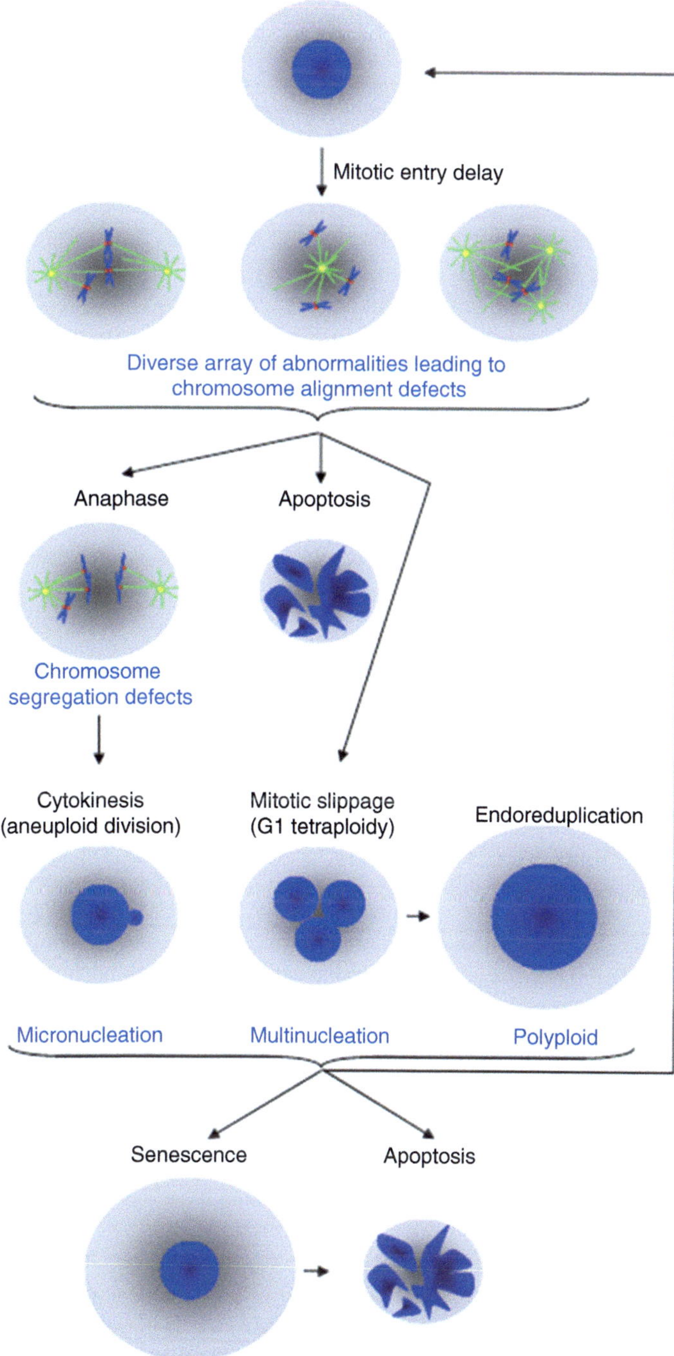

Fig. 2.2 Inhibition of the mitotic machinery can lead to a diverse array of outcomes. Several targets for antimitotic therapy participate in the transition from the G2 portion of the cell cycle to mitosis,

checkpoint. The spindle assembly checkpoint ensures that chromosomes are properly aligned to the metaphase plate prior to the anaphase initiation where sister chromatids segregate to opposite poles. Interestingly, at low concentrations of paclitaxel, inefficient chromosome alignment has been shown to occur without prolonged mitotic arrest, and the effect of paclitaxel is thus not dependent on its ability to induce mitotic arrest or delays (Chen and Horwitz 2002; Kelling et al. 2003).

For paclitaxel as well as its analog docetaxel, *in vitro* studies have demonstrated the presence of abnormal DNA contents and cell death even at concentrations where prolonged mitotic arrest does not occur (Chen and Horwitz 2002; Hernandez-Vargas et al. 2007a, b). Consistent with this finding, preclinical studies in xenograft models have failed to demonstrate a clear relationship between the degree of mitotic arrest and tumor growth inhibition (Gan et al. 1998; Milross et al. 1996; Schimming et al. 1999), and similar findings have been reported in a clinical setting (Symmans et al. 2000). This frustrating lack of a relationship between mitotic arrest and anticancer effect has represented somewhat of an obstacle for the rational development of antimitotic drugs, and clearly a more precise understanding of the means by which these drugs kill cells would facilitate their rational development.

How and why do antimitotic therapies elicit an antitumor response? The question has been surprisingly difficult to answer. Over the years, our understanding of the effects of antimitotic drugs has matured, with some surprises along the way. At this point, it has been well established that antimitotic compounds compromise the ability of cells to execute a successful division – cells will either fail to divide with a prolonged mitotic arrest that leads directly to cell death, or they divide abnormally, with an unequal distribution of DNA (Gascoigne and Taylor 2008; Rieder and Maiato 2004; Weaver and Cleveland 2005). Following such an unsuccessful division, cells may continue to cycle or undergo cell-cycle arrest or death. This diversity of outcomes following treatment with antimitotic agents has been shown to be dependent on cell type as well as on concentration of the antimitotic agent used (Gascoigne and Taylor 2008; Orth et al. 2008; Shi et al. 2008). Apoptosis has been shown to occur both during mitosis and in interphase following mitosis (Gascoigne and Taylor 2008; Shi et al. 2008). This may occur in part due to DNA double strand breaks that occur subsequent to treating cells with antimitotic agents (Dalton et al. 2007; Lei and Erikson 2008; Quignon et al. 2007). As apoptosis is not the only means of cell death in a solid tumor setting (Abend 2003), and forms of cell cycle arrest such as senescence contribute substantially to antitumor efficacy in preclinical models (Roninson et al. 2001), caution should be used in over interpreting switches toward and away from apoptosis as being indicators of drug sensitivity or resistance.

What implications do these mechanistic findings have for the rational development of antimitotic therapies? Clearly, the overreliance on the mitotic index as a means for optimizing drug development is one potential pitfall. The mitotic index is useful as a marker of drug effect, but more careful optimization of drugs in development can also be conducted by interrogating other effects of antimitotic agents that appear to be independent of mitotic arrest such as chromosome alignment or spindle bipolarity defects (Chakravarty et al. 2010). Another aspect of the complex biology of antimitotic agents is that there appears to be more flexibility in dosing these agents than was originally assumed. The prolonged mitotic arrest model suggests that sustained high concentrations of drug are required for antitumor effect. Findings with weekly taxane therapies, which have equivalent efficacy to once-every-three weeks taxane therapies, suggest that the same effect can be obtained by splitting the total dose of drug administered.

2.3   Next Generation Antimitotics

The Aurora kinases and Polo-like Kinases (PLKs) have obligate functions for normal cell cycle progression through mitosis. These kinases are the focus of intensive efforts by pharmaceutical companies as well as clinical and basic researchers for developing anticancer drugs. Therefore, these two kinase families are deserving of an in-depth description as examples of next-generation antimitotic targets.

2.3.1   Aurora Kinases

The Aurora kinases, first identified in yeast (Ipl1), Xenopus (Eg2), and Drosophila (Aurora), are critical regulators of mitosis (Andresson and Ruderman 1998; Chan and Botstein 1993; Glover et al. 1995; Roghi et al. 1998). In humans, three isoforms of Aurora kinase exist, including Aurora A, Aurora B, and Aurora C. Aurora A and Aurora B play critical roles in the normal progression of cells through mitosis, whereas Aurora C activity is largely restricted to meiotic cells.

Aurora C is predominantly expressed in testis (Tseng et al. 1998), though it has been detected in other cell types as well, including certain cancer cell lines (Kimura et al. 1999; Sasai et al. 2004; Takahashi et al. 2000). Forced overexpression of Aurora C in experimental models results in supernumerary centrosomes and polyploidy, and thus has been linked to oncogenic transformation (Dutertre et al. 2005). Despite these observations, however, it remains unclear if Aurora C functions in the mitotic division of somatic cells or in the natural history of cancer. Thus, this section will focus on the function of Aurora A and Aurora B in mitosis, their role in oncogenesis and on their utility as targets for cancer therapeutic intervention.

Aurora A and Aurora B are structurally closely related. Their catalytic domains lie in the C-terminus, where they differ in only a few amino acids. Greater diversity exists in their noncatalytic N-terminal domains. It is the sequence diversity in this region of Aurora A and Aurora B that dictates their interactions with distinct protein partners, allowing these kinases to have unique subcellular localizations and functions within mitotic cells. Consequently, attempts are in progress to develop small molecule inhibitor drugs targeting Aurora A, Aurora B, or both of these kinases simultaneously, as each of these approaches may provide unique modalities for the treatment of cancer.

The Aurora A gene (AURKA) localizes to chromosome 20q13.2, which is commonly amplified or overexpressed at a high incidence in a diverse array of tumor types (Bischoff et al. 1998; Camacho et al. 2006; Chng et al. 2006; Ikezoe et al. 2007; Sen et al. 2002). Increased Aurora A expression has been correlated to the etiology of cancer and to a worsened prognosis (Fraizer et al. 2004; Guan et al. 2007; Jeng et al. 2004; Landen et al. 2007; Miyoshi et al. 2001; Sakakura et al. 2001; Sen et al. 2002). This concept has been supported in experimental models, demonstrating that Aurora A overexpression leads to oncogenic transformation (Goepfert et al. 2002; Li et al. 2009; Wang et al. 2006a, b; Zhang et al. 2008; Zhou et al. 1998). Overexpression of Aurora A kinase is suspected to result in a stoichiometric imbalance between Aurora A and its regulatory partners, leading to chromosomal instability and subsequent transforming events. The potential oncogenic role of Aurora A has led to considerable interest in targeting this kinase for the treatment of cancer. However as Aurora A activity is requisite for normal mitotic progression, there is still no reason to suspect that cancers overexpressing Aurora A would be more or less sensitive to Aurora A targeted inhibition. Overexpression of Aurora B has also been reported in some cancers (Ikezoe et al. 2007). Similar to the case with Aurora A, overexpression of Aurora B has been correlated to a worsened prognosis in some cancers. In cases where either Aurora A or Aurora B have been demonstrated to be overexpressed, it is not always clear if the apparent overexpression is due to increased levels of protein per mitotic cell or more simply due to a higher mitotic index in some tumors.

During a normal cell cycle, Aurora A is first expressed in the G2 stage where it localizes to centrosomes and functions in centrosome maturation and separation as well as in the entry of cells into mitosis. Although Aurora A kinase inhibition results in a delayed mitotic entry (Marumoto et al. 2002), cells commonly enter mitosis despite having inactive Aurora A. In mitotic cells, Aurora A predominantly localizes to centrosomes and the proximal portion of incipient mitotic spindles. There it interacts with and phosphorylates a diverse set of proteins that collectively function in the formation of mitotic spindle poles and spindles, the attachment of spindles to sister chromatid at the kinetochores, the subsequent alignment and separation of chromosome, the spindle assembly checkpoint, and cytokinesis (Barr and Gergely 2007; Bischoff and Plowman 1999; Carmena and Earnshaw 2003; Giet et al. 2005).

The outcomes associated with inhibition of Aurora A have been studied using several experimental techniques; including gene mutation, RNA interference, antibody microinjection, and ATP-competitive small molecule kinase inhibitors (Glover et al. 1995; Hoar et al. 2007; Kaestner et al. 2009; Katayama et al. 2001; Marumoto et al. 2003; Sasai et al. 2008). Aurora A inhibition initially leads to the formation of abnormal mitotic spindles, either monopolar, bipolar, or tripolar with misaligned chromosomes, often accompanied by centrosome separation defects. These defects lead to a mitotic arrest, which presumably is mediated by activation of the spindle assembly checkpoint. The fate of these arrested cells can vary. In cases, prolonged mitotic arrest may lead directly to apoptosis. Some cells may also exit mitosis without undergoing cytokinesis resulting in G1 tetraploidy. Still further, cells may divide at a high frequency, albeit with severe chromosome segregation defects. In the latter two outcomes, the abnormal mitotic divisions can lead to deleterious aneuploidy resulting in cell death or arrest. This diversity in outcomes resulting from Aurora A kinase inhibition is shared by other antimitotic therapies (Gascoigne and Taylor 2008). Interestingly, the outcomes associated with Aurora A inhibition in many ways pheno-copy those associated with Aurora A overexpression, supporting the idea that there exist stoichiometric requirements on Aurora A for normal mitosis to occur.

Aurora B localizes to the centromeres in preanaphase cells. There it plays a critical role in spindle bipolarity and the establishment and maintenance of the spindle assembly checkpoint (Adams et al. 2001; Ditchfield et al. 2003; Hauf et al. 2003; Murata-Hori and Wang 2002). During anaphase and telophase, Aurora B localizes to the spindle midzone and midbody, respectively. There, Aurora B functions in cytokinesis (Giet and Glover 2001; Yokoyama et al. 2005). Inhibition of Aurora B through the use of gene mutations, RNA interference or ATP competitive small molecule inhibitors leads to defects in the attachment of the spindle microtubules to kinetochores, chromosome segregation, and formation of the cleavage furrow (Adams et al. 2001; Ditchfield et al. 2003; Giet and Glover 2001; Honda et al. 2003; Murata-Hori and Wang 2002; Yokoyama et al. 2005). Aurora B inhibition also prevents the proper formation of the spindle assembly checkpoint, causing cells to exit mitosis prematurely without a mitotic arrest and often without completing cytokinesis (Ditchfield et al. 2003; Hauf et al. 2003). The fate of these G1 tetraploid cells is to die, arrest, or undergo additional rounds of DNA replication (endoreduplication) resulting in a DNA ploidy of >4N.

Many small molecule inhibitors of the Aurora kinases have been or are being tested in clinical trials in cancer patients. This comprises Aurora A selective inhibitors, Aurora B selective inhibitors, or dual Aurora A and Aurora B inhibitors. Some of these inhibitors lack functional selectivity as they concurrently inhibit multiple kinases in addition to the Aurora kinases. In these cases, multiple mechanisms of actions may attribute to the pharmacodynamic and clinical activity as well as to the toxicities observed. As the Aurora kinases have obligate function in all dividing cells, these inhibitors are being developed in a diverse array of solid and hematological cancers, in both single agent and combination settings. Some evidence for single agent antitumor activity has been reported, including partial responses and prolonged stabilized disease in several solid tumor and hematological malignancy indications.

2.3.2 Polo-Like Kinases

The first PLK was identified in Drosophila melanogaster (polo), with orthologs also found in yeast (cdc5 and plo1) and Xenopus (Plx) (Kumagai and Dunphy 1996; Llamazares et al. 1991; Sunkel and Glover 1988). Each of these PLK orthologs are essential regulators of mitosis and are structurally and functionally related to the mammalian family member PLK1. The mammalian family is comprised of three additional members PLK2, PLK3, and PLK4. Like PLK1, PLK4 functions during mitosis, albeit in a different manner; PLK2 and PLK3 have nonmitotic roles in regulating the cell cycle (Winkles and Alberts 2005). Of the four mammalian PLK family members, PLK1 is the most extensively characterized and small molecule inhibitors developed against this isoform are being evaluated in preclinical and clinical settings for the treatment of cancer. Thus, this section will focus on the function of PLK1, its potential role in tumorigenesis, and its promise as a target for cancer therapy.

Several studies exemplify PLK1 as a compelling target for therapeutic intervention. Overexpression of PLK1 transforms cells such that they form tumors when grown as xenografts in immunocompromised mice (Smith et al. 1997). Strengthening the notion that PLK1 may contribute to the promotion and progression of cancers, PLK1 is overexpressed in a broad spectrum of solid and hematological malignancies and this overexpression is correlated with poor prognosis and survival in patients (Kneisel et al. 2002; Takai et al. 2001) (Dietzmann et al. 2001; Gray et al. 2004; Kanaji et al. 2006; Knecht et al. 1999; Mito et al. 2005; Takahashi et al. 2003; Tokumitsu et al. 1999; Yamamoto et al. 2006). To date, mutations or amplification of the PLK1 gene has not been detected.

The PLKs are highly conserved serine/threonine kinases distinguished by non-catalytic C-terminal domains of 60–70 amino acids termed the polo-box domain (PBD). The PBD serves as a binding module to phosphorylated motifs on other proteins mediating protein-protein interactions (Elia et al. 2003a, b; Lee et al. 1998). The kinase domain and PBD are thought to inhibit each other by intermolecular interaction during G1 and S phase, rendering the kinase inactive. Temporal control of PLK1 occurs during G2 by phosphorylation of the kinase domain, relieving interaction with the PBD. Cdk1 has emerged as a kinase that can phosphorylate proteins creating docking sites for the PBD of PLK1 (Fu et al. 2008; Neef et al. 2007; Wu et al. 2008). Spatial regulation of PLK1 occurs by the recruitment of the PBD to distinct mitotic locales enabling PLK1 to phosphorylate a variety of substrates that carry out divergent mitotic functions.

In G2, PLK1 localizes to centrosomes and redistributes elsewhere during mitosis. In metaphase, PLK1 is found at the centromeres and kinetochores, the spindle midzone in anaphase, and finally at the midbody during cytokinesis. PLK1 plays a role in regulating centrosome maturation, entry into mitosis, activity of the anaphase promoting complex, formation of and maintenance of a bipolar mitotic spindle, cytokinesis and mitotic exit (Eckerdt and Strebhardt 2006; Lane and Nigg 1996; Petronczki et al. 2007; Sumara et al. 2004; Toyoshima-Morimoto et al. 2001).

The consequences upon inhibition or downregulation of the protein have been studied by chemical and genetic tools, respectively (Lane and Nigg 1996; Lenart et al. 2007; Liu and Erikson 2002; Liu and Erikson 2003; Peters et al. 2006; Rudolph et al. 2009; Spankuch-Schmitt et al. 2002; Spankuch et al. 2004). Initial studies using small interfering RNA demonstrated that knockdown of PLK1 leads to prolonged mitotic delay and decreased cellular proliferation. Similar phenotypes are exhibited with small molecule inhibitors targeting the catalytic active site or those blocking the PBD in a broad range of tumor cell lines. Inhibition of PLK1 prevents localization at centrosomes and kinetochores, resulting in activation of the spindle assembly checkpoint. This manifests as a prometaphase mitotic delay characterized by monopolar or bipolar misaligned mitotic spindles that do not stably attach to kinetochores. Uniformly, studies have documented apoptosis as a consequence of this mitotic delay. Studies with a small molecule inhibitor also suggest that a cytostatic response results from the mitotic delay due to mitotic slippage (Gilmartin et al. 2009).

It has emerged that additional functions for PLK1 outside of mitosis exist. These include the possible involvement in the regulation of telomere stabilization, the regulation of DNA topoisomerase II, and DNA repair (Li et al. 2008; Svendsen et al. 2009). Activity of PLK1 is inhibited in the presence of DNA damage to ensure that these compromised cells do not progress into mitosis (Smits et al. 2000). However following satisfaction or relaxation of the DNA damage checkpoint, PLK1 is necessary to enable mitotic entry (van Vugt et al. 2004).

Small molecule inhibitors targeting the catalytic active site of PLK1 are under evaluation in clinical trials for both solid and hematological malignancies (Schoffski 2009). Clinical benefit has been observed for some tumor types in Phase I and has warranted Phase II studies for both single agent as well as combination trials.

2.4 Conclusion

Antimitotic approaches for therapeutic intervention of cancer have proven to be effective means for treating cancer. To date, these agents comprise the microtubule perturbing classes of molecules including the taxanes, the vinka alkaloids, and the epothilones. As the mechanism of action of these agents becomes clearer, more rational approaches for their clinical application as single agents or in combination with other therapeutics should emerge. Moreover, considerable efforts are ongoing to explore new modalities for perturbing the mitotic machinery by selectivity targeting key enzymatic mitotic regulators, for example the Aurora and PLKs. In early clinical testing, these agents have demonstrated promising activity, and molecules within these classes will likely emerge that provide improvements over current standard of care agents, including more manageable toxic side effects and improved responses in a distinct range of cancer indications alone or in combination with other therapeutic agents.

References

Abend M (2003) Reasons to reconsider the significance of apoptosis for cancer therapy. Int J Radiat Biol 79:927–941

Adams RR, Maiato H, Earnshaw WC, Carmena M (2001) Essential roles of Drosophila inner centromere protein (INCENP) and aurora B in histone H3 phosphorylation, metaphase chromosome alignment, kinetochore disjunction, and chromosome segregation. J Cell Biol 153:865–880

Andresson T, Ruderman JV (1998) The kinase Eg2 is a component of the Xenopus oocyte progesterone-activated signaling pathway. EMBO J 17:5627–5637

Barr AR, Gergely F (2007) Aurora-A: the maker and breaker of spindle poles. J Cell Sci 120:2987–2996

Bischoff JR, Plowman GD (1999) The Aurora/Ipl1p kinase family: regulators of chromosome segregation and cytokinesis. Trends Cell Biol 9:454–459

Bischoff JR, Anderson L, Zhu Y, Mossie K, Ng L, Souza B, Schryver B, Flanagan P, Clairvoyant F, Ginther C et al (1998) A homologue of Drosophila aurora kinase is oncogenic and amplified in human colorectal cancers. EMBO J 17:3052–3065

Bollag DM, McQueney PA, Zhu J, Hensens O, Koupal L, Liesch J, Goetz M, Lazarides E, Woods CM (1995) Epothilones, a new class of microtubule-stabilizing agents with a taxol-like mechanism of action. Cancer Res 55:2325–2333

Burstein HJ, Harris LN, Marcom PK, Lambert-Falls R, Havlin K, Overmoyer B, Friedlander RJ Jr, Gargiulo J, Strenger R, Vogel CL et al (2003) Trastuzumab and vinorelbine as first-line therapy for HER2-overexpressing metastatic breast cancer: multicenter phase II trial with clinical outcomes, analysis of serum tumor markers as predictive factors, and cardiac surveillance algorithm. J Clin Oncol 21:2889–2895

Camacho E, Bea S, Salaverria I, Lopez-Guillermo A, Puig X, Benavente Y, de Sanjose S, Campo E, Hernandez L (2006) Analysis of Aurora-A and hMPS1 mitotic kinases in mantle cell lymphoma. Int J Cancer 118:357–363

Carmena M, Earnshaw WC (2003) The cellular geography of aurora kinases. Nat Rev Mol Cell Biol 4:842–854

Chakravarty, A., Shinde, V., Galvin, K., Bowman, D., Stringer, B., Danaee, H., Venkatakrishnan, K., Simpson, C., Farron-Yowe, L., Liu, H., et al. (2010). Phase I Assessment of Mechanistic Pharmacodynamic Biomarkers for MLN8054, a Small-Molecule Inhibitor of Aurora A Kinase. Submitted for publication

Chan CS, Botstein D (1993) Isolation and characterization of chromosome-gain and increase-in-ploidy mutants in yeast. Genetics 135:677–691

Chen JG, Horwitz SB (2002) Differential mitotic responses to microtubule-stabilizing and -destabilizing drugs. Cancer Res 62:1935–1938

Chng WJ, Ahmann GJ, Henderson K, Santana-Davila R, Greipp PR, Gertz MA, Lacy MQ, Dispenzieri A, Kumar S, Rajkumar SV et al (2006) Clinical implication of centrosome amplification in plasma cell neoplasm. Blood 107:3669–3675

Dalton WB, Nandan MO, Moore RT, Yang VW (2007) Human cancer cells commonly acquire DNA damage during mitotic arrest. Cancer Res 67:11487–11492

Dietzmann K, Kirches E, Tighe A, von B, Jachau K, Mawrin C (2001) Increased human polo-like kinase-1 expression in gliomas. J Neurooncol 53:1–11

Ditchfield C, Johnson VL, Tighe A, Ellston R, Haworth C, Johnson T, Mortlock A, Keen N, Taylor SS (2003) Aurora B couples chromosome alignment with anaphase by targeting BubR1, Mad2, and Cenp-E to kinetochores. J Cell Biol 161:267–280

Dutertre S, Hamard-Peron E, Cremet JY, Thomas Y, Prigent C (2005) The absence of p53 aggravates polyploidy and centrosome number abnormality induced by Aurora-C overexpression. Cell Cycle 4:1783–1787

Eckerdt F, Strebhardt K (2006) Polo-like kinase 1: target and regulator of anaphase-promoting complex/cyclosome-dependent proteolysis. Cancer Res 66:6895–6898

Elia AE, Cantley LC, Yaffe MB (2003a) Proteomic screen finds pSer/pThr-binding domain localizing Plk1 to mitotic substrates. Science 299:1228–1231

Elia AE, Rellos P, Haire LF, Chao JW, Ivins FJ, Hoepker K, Mohammad D, Cantley LC, Smerdon SJ, Yaffe MB (2003b) The molecular basis for phosphodependent substrate targeting and regulation of Plks by the Polo-box domain. Cell 115:83–95

Fraizer GC, Diaz MF, Lee IL, Grossman HB, Sen S (2004) Aurora-A/STK15/BTAK enhances chromosomal instability in bladder cancer cells. Int J Oncol 25:1631–1639

Fu Z, Malureanu L, Huang J, Wang W, Li H, van Deursen JM, Tindal DJ, Chen J (2008) Plk1-dependent phosphorylation of FoxM1 regulates a transcriptional programme required for mitotic progression. Nat Cell Biol 10:1076–1082

Fumoleau P, Coudert B, Isambert N, Ferrant E (2007) Novel tubulin-targeting agents: anticancer activity and pharmacologic profile of epothilones and related analogues. Ann Oncol 18(Suppl 5):v9–v15

Gan Y, Wientjes MG, Lu J, Au JL (1998) Cytostatic and apoptotic effects of paclitaxel in human breast tumors. Cancer Chemother Pharmacol 42:177–182

Gascoigne KE, Taylor SS (2008) Cancer cells display profound intra- and interline variation following prolonged exposure to antimitotic drugs. Cancer Cell 14:111–122

Giet R, Glover DM (2001) Drosophila aurora B kinase is required for histone H3 phosphorylation and condensin recruitment during chromosome condensation and to organize the central spindle during cytokinesis. J Cell Biol 152:669–682

Giet R, Petretti C, Prigent C (2005) Aurora kinases, aneuploidy and cancer, a coincidence or a real link? Trends Cell Biol 15:241–250

Gilmartin AG, Bleam MR, Richter MC, Erskine SG, Kruger RG, Madden L, Hassler DF, Smith GK, Gontarek RR, Courtney MP, Sutton D, Diamond MA, Jackson JR, Laquerre SG (2009) Distinct concentration-dependent effects of the polo-like kinase 1-specific inhibitor GSK461364A, including differential effect on apoptosis. Cancer Res 69:6969–6977

Glover DM, Leibowitz MH, McLean DA, Parry H (1995) Mutations in aurora prevent centrosome separation leading to the formation of monopolar spindles. Cell 81:95–105

Goepfert TM, Adigun YE, Zhong L, Gay J, Medina D, Brinkley WR (2002) Centrosome amplification and overexpression of aurora A are early events in rat mammary carcinogenesis. Cancer Res 62:4115–4122

Gonzalez-Angulo AM, Hortobagyi GN (2008) Optimal schedule of paclitaxel: weekly is better. J Clin Oncol 26:1585–1587

Gray PJ Jr, Bearss DJ, Han H, Nagle R, Tsao MS, Dean N, Von Hoff DD (2004) Identification of human polo-like kinase 1 as a potential therapeutic target in pancreatic cancer. Mol Cancer Ther 3:641–646

Guan Z, Wang XR, Zhu XF, Huang XF, Xu J, Wang LH, Wan XB, Long ZJ, Liu JN, Feng GK et al (2007) Aurora-A, a negative prognostic marker, increases migration and decreases radiosensitivity in cancer cells. Cancer Res 67:10436–10444

Harrison MR, Holen KD, Liu G (2009) Beyond taxanes: a review of novel agents that target mitotic tubulin and microtubules, kinases, and kinesins. Clin Adv Hematol Oncol 7:54–64

Hauf S, Cole RW, LaTerra S, Zimmer C, Schnapp G, Walter R, Heckel A, van Meel J, Rieder CL, Peters JM (2003) The small molecule Hesperadin reveals a role for Aurora B in correcting kinetochore-microtubule attachment and in maintaining the spindle assembly checkpoint. J Cell Biol 161:281–294

Hernandez-Vargas H, Palacios J, Moreno-Bueno G (2007a) Telling cells how to die: docetaxel therapy in cancer cell lines. Cell Cycle 6:780–783

Hernandez-Vargas H, von Kobbe C, Sanchez-Estevez C, Julian-Tendero M, Palacios J, Moreno-Bueno G (2007b) Inhibition of paclitaxel-induced proteasome activation influences paclitaxel cytotoxicity in breast cancer cells in a sequence-dependent manner. Cell Cycle 6:2662–2668

Hoar K, Chakravarty A, Rabino C, Wysong D, Bowman D, Roy D'Amore N, Ecsedy JA (2007) MLN8054, a small molecule inhibitor of Aurora A, causes spindle pole and chromosome congression defects leading to aneuploidy. Mol Cell Biol 27:4513–4525

Honda R, Korner R, Nigg EA (2003) Exploring the functional interactions between Aurora B, INCENP, and survivin in mitosis. Mol Biol Cell 14:3325–3341

Ibrahim NK, Desai N, Legha S, Soon-Shiong P, Theriault RL, Rivera E, Esmaeli B, Ring SE, Bedikian A, Hortobagyi GN, Ellerhorst JA (2002) Phase I and pharmacokinetic study of ABI-007, a Cremophor-free, protein-stabilized, nanoparticle formulation of paclitaxel. Clin Cancer Res 8:1038–1044

Ikezoe T, Yang J, Nishioka C, Tasaka T, Taniguchi A, Kuwayama Y, Komatsu N, Bandobashi K, Togitani K, Koeffler HP, Taguchi H (2007) A novel treatment strategy targeting Aurora kinases in acute myelogenous leukemia. Mol Cancer Ther 6:1851–1857

Jahanzeb M, Mortimer JE, Yunus F, Irwin DH, Speyer J, Koletsky AJ, Klein P, Sabir T, Kronish L (2002) Phase II trial of weekly vinorelbine and trastuzumab as first-line therapy in patients with HER2(+) metastatic breast cancer. Oncologist 7:410–417

Jeng YM, Peng SY, Lin CY, Hsu HC (2004) Overexpression and amplification of Aurora-A in hepatocellular carcinoma. Clin Cancer Res 10:2065–2071

Jones SE, Erban J, Overmoyer B, Budd GT, Hutchins L, Lower E, Laufman L, Sundaram S, Urba WJ, Pritchard KI et al (2005) Randomized phase III study of docetaxel compared with paclitaxel in metastatic breast cancer. J Clin Oncol 23:5542–5551

Kaestner P, Stolz A, Bastians H (2009) Determinants for the efficiency of anticancer drugs targeting either Aurora-A or Aurora-B kinases in human colon carcinoma cells. Mol Cancer Ther 8:2046–2056

Kanaji S, Saito H, Tsujitani S, Matsumoto S, Tatebe S, Kondo A, Ozaki M, Ito H, Ikeguchi M (2006) Expression of polo-like kinase 1 (PLK1) protein predicts the survival of patients with gastric carcinoma. Oncology 70:126–133

Katayama H, Zhou H, Li Q, Tatsuka M, Sen S (2001) Interaction and feedback regulation between STK15/BTAK/Aurora-A kinase and protein phosphatase 1 through mitotic cell division cycle. J Biol Chem 276:46219–46224

Kelling J, Sullivan K, Wilson L, Jordan MA (2003) Suppression of centromere dynamics by Taxol in living osteosarcoma cells. Cancer Res 63:2794–2801

Kimura M, Matsuda Y, Yoshioka T, Okano Y (1999) Cell cycle-dependent expression and centrosome localization of a third human aurora/Ipl1-related protein kinase, AIK3. J Biol Chem 274:7334–7340

Knecht R, Elez R, Oechler M, Solbach C, von Ilberg C, Strebhardt K (1999) Prognostic significance of polo-like kinase (PLK) expression in squamous cell carcinomas of the head and neck. Cancer Res 59:2794–2797

Kneisel L, Strebhardt K, Bernd A, Wolter M, Binder A, Kaufmann R (2002) Expression of polo-like kinase (PLK1) in thin melanomas: a novel marker of metastatic disease. J Cutan Pathol 29:354–358

Kowalski RJ, Giannakakou P, Hamel E (1997) Activities of the microtubule-stabilizing agents epothilones A and B with purified tubulin and in cells resistant to paclitaxel (Taxol(R)). J Biol Chem 272:2534–2541

Krikorian A, Breillout F (1991) Vinorelbine (Navelbine). A new semisynthetic vinca alkaloid. Onkologie 14:7–12

Kumagai A, Dunphy WG (1996) Purification and molecular cloning of Plx1, a Cdc25-regulatory kinase from Xenopus egg extracts. Science 273:1377–1380

Landen CN Jr, Lin YG, Immaneni A, Deavers MT, Merritt WM, Spannuth WA, Bodurka DC, Gershenson DM, Brinkley WR, Sood AK (2007) Overexpression of the centrosomal protein Aurora-A kinase is associated with poor prognosis in epithelial ovarian cancer patients. Clin Cancer Res 13:4098–4104

Lane HA, Nigg EA (1996) Antibody microinjection reveals an essential role for human polo-like kinase 1 (Plk1) in the functional maturation of mitotic centrosomes. J Cell Biol 135:1701–1713

Lee KS, Grenfell TZ, Yarm FR, Erikson RL (1998) Mutation of the polo-box disrupts localization and mitotic functions of the mammalian polo kinase Plk. Proc Natl Acad Sci USA 95:9301–9306

Lei M, Erikson RL (2008) Plk1 depletion in nontransformed diploid cells activates the DNA-damage checkpoint. Oncogene 27:3935–3943

Lenart P, Petronczki M, Steegmaier M, Di Fiore B, Lipp JJ, Hoffmann M, Rettig WJ, Kraut N, Peters JM (2007) The small-molecule inhibitor BI 2536 reveals novel insights into mitotic roles of polo-like kinase 1. Curr Biol 17:304–315

Li H, Wang Y, Liu X (2008) Plk1-dependent phosphorylation regulates functions of DNA topoisomerase IIalpha in cell cycle progression. J Biol Chem 283:6209–6221

Li CC, Chu HY, Yang CW, Chou CK, Tsai TF (2009) Aurora-A overexpression in mouse liver causes p53-dependent premitotic arrest during liver regeneration. Mol Cancer Res 7:678–688

Liu X, Erikson RL (2002) Activation of Cdc2/cyclin B and inhibition of centrosome amplification in cells depleted of Plk1 by siRNA. Proc Natl Acad Sci USA 99:8672–8676

Liu X, Erikson RL (2003) Polo-like kinase (Plk)1 depletion induces apoptosis in cancer cells. Proc Natl Acad Sci USA 100:5789–5794

Llamazares S, Moreira A, Tavares A, Girdham C, Spruce BA, Gonzalez C, Karess RE, Glover DM, Sunkel CE (1991) Polo encodes a protein kinase homolog required for mitosis in Drosophila. Genes Dev 5:2153–2165

Marumoto T, Hirota T, Morisaki T, Kunitoku N, Zhang D, Ichikawa Y, Sasayama T, Kuninaka S, Mimori T, Tamaki N et al (2002) Roles of aurora-A kinase in mitotic entry and G2 checkpoint in mammalian cells. Genes Cells 7:1173–1182

Marumoto T, Honda S, Hara T, Nitta M, Hirota T, Kohmura E, Saya H (2003) Aurora-A kinase maintains the fidelity of early and late mitotic events in HeLa cells. J Biol Chem 278: 51786–51795

Milross CG, Mason KA, Hunter NR, Chung WK, Peters LJ, Milas L (1996) Relationship of mitotic arrest and apoptosis to antitumor effect of paclitaxel. J Natl Cancer Inst 88:1308–1314

Mito K, Kashima K, Kikuchi H, Daa T, Nakayama I, Yokoyama S (2005) Expression of Polo-Like Kinase (PLK1) in non-Hodgkin's lymphomas. Leuk Lymphoma 46:225–231

Miyoshi Y, Iwao K, Egawa C, Noguchi S (2001) Association of centrosomal kinase STK15/BTAK mRNA expression with chromosomal instability in human breast cancers. Int J Cancer 92:370–373

Murata-Hori M, Wang YL (2002) The kinase activity of aurora B is required for kinetochore-microtubule interactions during mitosis. Curr Biol 12:894–899

Neef R, Gruneberg U, Kopajtich R, Li X, Nigg EA, Sillje H, Barr FA (2007) Choice of Plk1 docking partners during mitosis and cytokinesis is controlled by the activation state of Cdk1. Nat Cell Biol 9:436–444

Orth JD, Tang Y, Shi J, Loy CT, Amendt C, Wilm C, Zenke FT, Mitchison TJ (2008) Quantitative live imaging of cancer and normal cells treated with Kinesin-5 inhibitors indicates significant differences in phenotypic responses and cell fate. Mol Cancer Ther 7:3480–3489

Perez EA, Lerzo G, Pivot X, Thomas E, Vahdat L, Bosserman L, Viens P, Cai C, Mullaney B, Peck R, Hortobagyi GN (2007) Efficacy and safety of ixabepilone (BMS-247550) in a phase II study of patients with advanced breast cancer resistant to an anthracycline, a taxane, and capecitabine. J Clin Oncol 25:3407–3414

Peters U, Cherian J, Kim JH, Kwok BH, Kapoor TM (2006) Probing cell-division phenotype space and Polo-like kinase function using small molecules. Nat Chem Biol 2:618–626

Petronczki M, Glotzer M, Kraut N, Peters JM (2007) Polo-like kinase 1 triggers the initiation of cytokinesis in human cells by promoting recruitment of the RhoGEF Ect2 to the central spindle. Dev Cell 12:713–725

Quignon F, Rozier L, Lachages AM, Bieth A, Simili M, Debatisse M (2007) Sustained mitotic block elicits DNA breaks: one-step alteration of ploidy and chromosome integrity in mammalian cells. Oncogene 26:165–172

Rieder CL, Maiato H (2004) Stuck in division or passing through: what happens when cells cannot satisfy the spindle assembly checkpoint. Dev Cell 7:637–651

Roghi C, Giet R, Uzbekov R, Morin N, Chartrain I, Le Guellec R, Couturier A, Doree M, Philippe M, Prigent C (1998) The Xenopus protein kinase pEg2 associates with the centrosome in a cell cycle-dependent manner, binds to the spindle microtubules and is involved in bipolar mitotic spindle assembly. J Cell Sci 111(Pt 5):557–572

Roninson IB, Broude EV, Chang BD (2001) If not apoptosis, then what? Treatment-induced senescence and mitotic catastrophe in tumor cells. Drug Resist Updat 4:303–313

Rubin EH, Rothermel J, Tesfaye F, Chen T, Hubert M, Ho YY, Hsu CH, Oza AM (2005) Phase I dose-finding study of weekly single-agent patupilone in patients with advanced solid tumors. J Clin Oncol 23:9120–9129

Rudolph D, Steegmaier M, Hoffmann M, Grauert M, Baum A, Quant J, Haslinger C, Garin-Chesa P, Adolf GR (2009) BI 6727, a Polo-like kinase inhibitor with improved pharmacokinetic profile and broad antitumor activity. Clin Cancer Res 15:3094–3102

Sakakura C, Hagiwara A, Yasuoka R, Fujita Y, Nakanishi M, Masuda K, Shimomura K, Nakamura Y, Inazawa J, Abe T, Yamagishi H (2001) Tumour-amplified kinase BTAK is amplified and overexpressed in gastric cancers with possible involvement in aneuploid formation. Br J Cancer 84:824–831

Salminen E, Bergman M, Huhtala S, Ekholm E (1999) Docetaxel: standard recommended dose of 100 mg/m(2) is effective but not feasible for some metastatic breast cancer patients heavily pretreated with chemotherapy-A phase II single-center study. J Clin Oncol 17:1127

Sasai K, Katayama H, Stenoien DL, Fujii S, Honda R, Kimura M, Okano Y, Tatsuka M, Suzuki F, Nigg EA et al (2004) Aurora-C kinase is a novel chromosomal passenger protein that can complement Aurora-B kinase function in mitotic cells. Cell Motil Cytoskeleton 59:249–263

Sasai K, Parant JM, Brandt ME, Carter J, Adams HP, Stass SA, Killary AM, Katayama H, Sen S (2008) Targeted disruption of Aurora A causes abnormal mitotic spindle assembly, chromosome misalignment and embryonic lethality. Oncogene 27:4122–4127

Schiff PB, Horwitz SB (1980) Taxol stabilizes microtubules in mouse fibroblast cells. Proc Natl Acad Sci USA 77:1561–1565

Schimming R, Mason KA, Hunter N, Weil M, Kishi K, Milas L (1999) Lack of correlation between mitotic arrest or apoptosis and antitumor effect of docetaxel. Cancer Chemother Pharmacol 43:165–172

Schoffski P (2009) Polo-like kinase (PLK) inhibitors in preclinical and early clinical development in oncology. Oncologist 14:559–570

Seidman AD, Berry D, Cirrincione C, Harris L, Muss H, Marcom PK, Gipson G, Burstein H, Lake D, Shapiro CL et al (2008) Randomized phase III trial of weekly compared with every-3-weeks paclitaxel for metastatic breast cancer, with trastuzumab for all HER-2 overexpressors and random assignment to trastuzumab or not in HER-2 nonoverexpressors: final results of Cancer and Leukemia Group B protocol 9840. J Clin Oncol 26:1642–1649

Sen S, Zhou H, Zhang RD, Yoon DS, Vakar-Lopez F, Ito S, Jiang F, Johnston D, Grossman HB, Ruifrok AC et al (2002) Amplification/overexpression of a mitotic kinase gene in human bladder cancer. J Natl Cancer Inst 94:1320–1329

Sessa C, Perotti A, Llado A, Cresta S, Capri G, Voi M, Marsoni S, Corradino I, Gianni L (2007) Phase I clinical study of the novel epothilone B analogue BMS-310705 given on a weekly schedule. Ann Oncol 18:1548–1553

Shi J, Orth JD, Mitchison T (2008) Cell type variation in responses to antimitotic drugs that target microtubules and kinesin-5. Cancer Res 68:3269–3276

Smith MR, Wilson ML, Hamanaka R, Chase D, Kung H, Longo DL, Ferris DK (1997) Malignant transformation of mammalian cells initiated by constitutive expression of the polo-like kinase. Biochem Biophys Res Commun 234:397–405

Smits VA, Klompmaker R, Arnaud L, Rijksen G, Nigg EA, Medema RH (2000) Polo-like kinase-1 is a target of the DNA damage checkpoint. Nat Cell Biol 2:672–676

Spankuch B, Matthess Y, Knecht R, Zimmer B, Kaufmann M, Strebhardt K (2004) Cancer inhibition in nude mice after systemic application of U6 promoter-driven short hairpin RNAs against PLK1. J Natl Cancer Inst 96:862–872

Spankuch-Schmitt B, Bereiter-Hahn J, Kaufmann M, Strebhardt K (2002) Effect of RNA silencing of polo-like kinase-1 (PLK1) on apoptosis and spindle formation in human cancer cells. J Natl Cancer Inst 94:1863–1877

Sumara I, Gimenez-Abian JF, Gerlich D, Hirota T, Kraft C, de la Torre C, Ellenberg J, Peters JM (2004) Roles of polo-like kinase 1 in the assembly of functional mitotic spindles. Curr Biol 14:1712–1722

Sunkel CE, Glover DM (1988) Polo, a mitotic mutant of Drosophila displaying abnormal spindle poles. J Cell Sci 89:25–38

Svendsen JM, Smogorzewska A, Sowa ME, O'Connell BC, Gygi SP, Elledge SJ, Harper JW (2009) Mammalian BTBD12/SLX4 assembles a Holliday junction resolvase and is required for DNA repair. Cell 138:63–77

Symmans WF, Volm MD, Shapiro RL, Perkins AB, Kim AY, Demaria S, Yee HT, McMullen H, Oratz R, Klein P et al (2000) Paclitaxel-induced apoptosis and mitotic arrest assessed by serial fine-needle aspiration: implications for early prediction of breast cancer response to neoadjuvant treatment. Clin Cancer Res 6:4610–4617

Takahashi T, Futamura M, Yoshimi N, Sano J, Katada M, Takagi Y, Kimura M, Yoshioka T, Okano Y, Saji S (2000) Centrosomal kinases, HsAIRK1 and HsAIRK3, are overexpressed in primary colorectal cancers. Jpn J Cancer Res 91:1007–1014

Takahashi T, Sano B, Nagata T, Kato H, Sugiyama Y, Kunieda K, Kimura M, Okano Y, Saji S (2003) Polo-like kinase 1 (PLK1) is overexpressed in primary colorectal cancers. Cancer Sci 94:148–152

Takai N, Miyazaki T, Fujisawa K, Nasu K, Hamanaka R, Miyakawa I (2001) Polo-like kinase (PLK) expression in endometrial carcinoma. Cancer Lett 169:41–49

Thomas ES, Gomez HL, Li RK, Chung HC, Fein LE, Chan VF, Jassem J, Pivot XB, Klimovsky JV, de Mendoza FH et al (2007) Ixabepilone plus capecitabine for metastatic breast cancer progressing after anthracycline and taxane treatment. J Clin Oncol 25:5210–5217

Tokumitsu Y, Mori M, Tanaka S, Akazawa K, Nakano S, Niho Y (1999) Prognostic significance of polo-like kinase expression in esophageal carcinoma. Int J Oncol 15:687–692

Toyoshima-Morimoto F, Taniguchi E, Shinya N, Iwamatsu A, Nishida E (2001) Polo-like kinase 1 phosphorylates cyclin B1 and targets it to the nucleus during prophase. Nature 410:215–220 [erratum appears in Nature, 2001 Apr 12;410(6830):847]

Tseng TC, Chen SH, Hsu YP, Tang TK (1998) Protein kinase profile of sperm and eggs: cloning and characterization of two novel testis-specific protein kinases (AIE1, AIE2) related to yeast and fly chromosome segregation regulators. DNA Cell Biol 17:823–833

van Vugt MA, Bras A, Medema RH (2004) Polo-like kinase-1 controls recovery from a G2 DNA damage-induced arrest in mammalian cells. Mol Cell 15:799–811

Wang X, Zhou YX, Qiao W, Tominaga Y, Ouchi M, Ouchi T, Deng CX (2006a) Overexpression of aurora kinase A in mouse mammary epithelium induces genetic instability preceding mammary tumor formation. Oncogene 25:7148–7158

Wang XX, Liu R, Jin SQ, Fan FY, Zhan QM (2006b) Overexpression of Aurora-A kinase promotes tumor cell proliferation and inhibits apoptosis in esophageal squamous cell carcinoma cell line. Cell Res 16:356–366

Wartmann M, Altmann KH (2002) The biology and medicinal chemistry of epothilones. Curr Med Chem 2:123–148

Watkins EB, Chittiboyina AG, Jung JC, Avery MA (2005) The epothilones and related analogues-a review of their syntheses and anti-cancer activities. Curr Pharm Des 11:1615–1653

Weaver BAA, Cleveland DW (2005) Decoding the links between mitosis, cancer, and chemotherapy: the mitotic checkpoint, adaptation and cell death. Cancer Cell 8:7–12

Winkles JA, Alberts GF (2005) Differential regulation of polo-like kinase 1, 2, 3, and 4 gene expression in mammalian cells and tissues. Oncogene 24:260–266

Wu ZQ, Yang X, Weber G, Liu X (2008) Plk1 phosphorylation of TRF1 is essential for its binding to telomeres. J Biol Chem 283:25503–25513

Yamamoto Y, Matsuyama H, Kawauchi S, Matsumoto H, Nagao K, Ohmi C, Sakano S, Furuya T, Oga A, Naito K, Sasaki K (2006) Overexpression of polo-like kinase 1 (PLK1) and chromosomal instability in bladder cancer. Oncology 70:231–237

Yokoyama T, Goto H, Izawa I, Mizutani H, Inagaki M (2005) Aurora-B and Rho-kinase/ROCK, the two cleavage furrow kinases, independently regulate the progression of cytokinesis: possible existence of a novel cleavage furrow kinase phosphorylates ezrin/radixin/moesin (ERM). Genes Cells 10:127–137

Zhang D, Shimizu T, Araki N, Hirota T, Yoshie M, Ogawa K, Nakagata N, Takeya M, Saya H (2008) Aurora A overexpression induces cellular senescence in mammary gland hyperplastic tumors developed in p53-deficient mice. Oncogene 27:4305–4314

Zhou H, Kuang J, Zhong L, Kuo WL, Gray JW, Sahin A, Brinkley BR, Sen S (1998) Tumour amplified kinase STK15/BTAK induces centrosome amplification, aneuploidy and transformation. Nat Genet 20:189–193

Chapter 3
DNA Damage Checkpoint Signaling Pathways in Human Cancer

Robert T. Abraham and Thanos D. Halazonetis

3.1 Introduction

The genomic integrity of all organisms is constantly challenged by DNA damaging agents and errors that occur during normal physiological processes, such as DNA replication. In response to these challenges, DNA damage checkpoint pathways have evolved (Hartwell & Weinert 1989; Jackson & Bartek 2009; Osborn et al. 2002). These pathways utilize multiple proteins that perform functions ranging from sensing the presence of DNA damage to signaling to effectors that regulate cell cycle progression, cell survival, and repair.

In earlier days, the study of DNA damage checkpoint pathways focused on understanding the response of cells to ionizing radiation. For this reason genes that function in these pathways were called RAD followed by a number that usually reflects their order of discovery (Friedberg 1991; Ivanov & Haber 1997). With the demonstration that cancer cells have genetic mutations, which are responsible for the transformed phenotype (Parada et al. 1982; Bishop 1991), it became apparent that aberrant function of RAD genes could lead to genomic mutations that are responsible for cancer development (Loeb 1991; Kinzler & Vogelstein 1997; Paulovich et al. 1997). Such mutations could activate proto-oncogenes, converting them to oncogenes that promote cancer development, or could inactivate tumor suppressor genes, whose normal function is to curtail tumor development (Klein 1987; Levine 1990).

R.T. Abraham (⊠)
Center for Integrative Biology and Biotherapeutics, Pfizer Biopharmaceuticals,
401 N. Middletown Road, Pearl River, NY 10965, USA
e-mail: Robert.Abraham@pfizer.com

T.D. Halazonetis
Department of Molecular Biology and Biochemistry, University of Geneva,
Geneva CH-1205, Switzerland

D.A. Frank (ed.), *Signaling Pathways in Cancer Pathogenesis and Therapy*,
DOI 10.1007/978-1-4614-1216-8_3, © Springer Science+Business Media, LLC 2012

More recently, it has been shown that, in precancerous lesions and cancers, oncogenes induce DNA damage that leads to activation of the DNA damage checkpoint pathways (Bartkova et al. 2005; Gorgoulis et al. 2005). Since activation of these pathways can lead to cell cycle arrest and/or apoptosis (Hartwell & Weinert 1989; Jackson & Bartek 2009; Osborn et al. 2002; Clarke et al. 1993; Lowe et al. 1993; Di Leonardo et al. 1994), it is not surprising that cancer progression is associated with mutations targeting key players of these pathways (Hollstein et al. 1991; Sjoblom et al. 2006; Wood et al. 2007; Jones et al. 2008; Parsons et al. 2008; Cancer Genome Atlas Research Network 2008; Ding et al. 2008).

In this chapter, we will review the key DNA damage signaling checkpoint pathways that appear relevant to cancer development. We note that it is impossible to cover every aspect of this field. Further, although much of the knowledge about these pathways has been gleaned from the study of yeast and other model organisms, we will focus our review on higher eukaryotes and, where possible, on humans and mice, given the focus of this volume on cancer.

3.2 The Response to DNA Double Strand Breaks

From the perspective of human cancer, of the many pathways that respond to DNA damage, the pathways that respond to DNA double strand breaks (DSBs) and to stalled/collapsed DNA replication forks are the most relevant. In this section we will review the pathway that responds to DNA DSBs (Fig. 3.1).

Key proteins involved in the response to DNA DSBs are the kinases ataxia-telangiectasia mutated (ATM) and Chk2; the Mre11-Rad50-Nbs1 nuclease complex; the proteins Sae2/Ctp1/CtIP, p53 binding protein 1 (53BP1), mediator of the DNA damage checkpoint 1 (MDC1), RAP80, and Abraxas; the ubiquitin ligases RNF8, RNF168, and breast cancer susceptibility 1 (BRCA1); the histone H2AX; and the p53 tumor suppressor protein (Jackson & Bartek 2009; Osborn et al. 2002).

ATM is a large protein kinase belonging to the family of PI3K-related kinases (PIKKs) (Keith & Schreiber 1995; Manning et al. 2002; Shiloh 2003). It is believed to consist mostly of helical repeats packing one against the other to form a large arc-like structure (Perry & Kleckner 2003). The kinase domain resides at the very C-terminus of the protein. In the absence of DNA damage, ATM forms homodimers that lack kinase activity. After induction of DNA DSBs, ATM dissociates into monomers and becomes autophosphorylated on Ser1981, a residue within the helical repeat region (Bakkenist & Kastan 2003). Ser1981 phosphorylation is a very good marker of activated ATM; this phosphorylation may also help determine the equilibrium between active and inactive ATM (Daniel et al. 2008).

The Mre11-Rad50-Nbs1 (MRN) complex assembles as a hetero-hexamer consisting of two Mre11, two Rad50, and two Nbs1 subunits (Williams et al. 2007). Mre11 is a nuclease that trims the DNA ends and possesses both endonuclease activity for single-stranded DNA and 3′-5′ exonuclease activity for double-stranded DNA (Williams et al. 2008). Mre11 interacts with Rad50, a protein that contains a

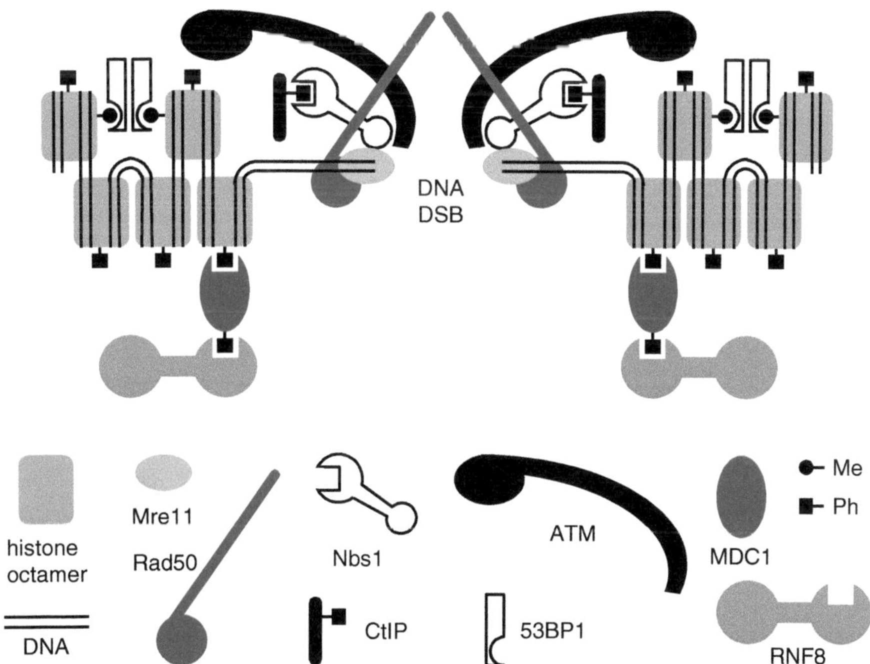

Fig. 3.1 DNA double-strand break (DSB) signaling pathways. The Mre11-Rad50-Nbs1 complex is recruited at the DNA ends. This complex then recruits ATM and CtIP. ATM phosphorylates histone H2AX leading to the sequential recruitment of MDC1 and RNF8. In parallel, unstacking of nucleosomes exposes binding sites for 53BP1. See text for more details. *Me* methylation; *Ph* phosphorylation

bipartite ATPase domain (related to the ATPase domain present in ABC transporters) and a long helical coil separating the N- and C-terminal parts of the ATPase domain (Hopfner et al. 2000). At its center, the coiled coil is interrupted by a small zinc-binding domain that induces a 180° turn in the direction of the polypeptide, thus allowing the N-terminal and C-terminal parts of the ATPase domain to interact with one another (Hopfner et al. 2002). The ATPase domain of Rad50 is physically next to the Mre11 nuclease domain and together these two domains bind DNA in an ATP-dependent manner (Hopfner et al. 2001). Since the MRN complex exists as a dimer, it is capable of holding together the two DNA ends at DSB sites.

Nbs1, the third protein of the MRN complex, contains two small C-terminal motifs: one of these binds to Mre11 and the other to ATM (Falck et al. 2005). The N-terminus of Nbs1 contains an FHA domain fused to two BRCT domains. These three phosphopeptide binding domains mediate interactions of Nbs1 with at least two proteins: Sae2/Ctp1/CtIP (via the FHA domain) and MDC1 (probably via both the FHA and BRCT domains) (Williams et al. 2009; Lloyd et al. 2009). Sae2/Ctp1/CtIP regulates the nuclease activity of Mre11 and facilitates limited resection of one of the strands of the DNA to create a single-stranded DNA overhang that can be used for DNA repair (Sartori et al. 2007). The MRN complex is believed to be one of the first protein complexes to be recruited to sites of DNA DSBs (Adelman & Petrini 2009). It is important for both DNA repair, by holding the DNA ends together and by processing the DNA ends, and for activation of the checkpoint by recruiting and activating ATM (Falck et al. 2005; Morales et al. 2005; Lee & Paull 2005).

Histone H2AX is a variant of histone H2A that contains at its C-terminus a consensus site for ATM-dependent phosphorylation. About 10% of histone H2A in the genome is the H2AX variant (Redon et al. 2002). After formation of a DNA DSB, H2AX in the vicinity of the break becomes phosphorylated in an ATM-dependent manner (Rogakou et al. 1998). This phosphorylation may extend quite far from the break corresponding to a DNA length of about 1 Mbp (about 5,000 nucleosomes) (Rogakou et al. 1999). The phosphorylated H2AX C-terminus serves as a platform for recruitment of DNA damage response proteins, most notably of MDC1 (Stucki et al. 2005).

MDC1 is a protein that contains at its C-terminus a pair of BRCT domains (Goldberg et al. 2003). This pair of BRCT domains interacts in a phospho-dependent manner with the phosphorylated C-terminus of histone H2AX (Stucki et al. 2005). Once recruited to the chromatin flanking DNA DSBs, MDC1 becomes phosphory-lated at its N-terminus by ATM, creating binding sites for additional DNA damage response proteins, such as Nbs1 and RNF8 (Lloyd et al. 2009; Mailand et al. 2007). Nbs1 was discussed above in the context of the MRN complex. The interaction of Nbs1 with MDC1 provides a second mechanism for recruitment of the MRN com-plex to sites of DNA DSBs far away from the DNA end. The significance of this recruitment is unclear, but may serve to provide better cohesion between the DNA molecules and higher levels of ATM activation.

RNF8 is also recruited to sites of DNA DSBs by interacting with MDC1 (Mailand et al. 2007). Its recruitment is dependent on interactions of its N-terminal FHA domain with sites on MDC1 that become phosphorylated by ATM. RNF8 is a ubiq-uitin protein ligase and this activity is mediated by its C-terminal RING domain. It has been proposed that histone H2A is the target of RNF8-dependent ubiquitylation (Mailand et al. 2007; Huen et al. 2007). H2A ubiquitylation facilitates the recruit-ment of additional DNA damage response proteins that function in both checkpoint activation and DNA repair. These proteins include RNF168, another ubiquitin ligase, that further amplifies protein ubiquitylation at sites of DNA DSBs (Doil et al. 2009; Stewart et al. 2009), and RAP80 and Abraxas, which recruit BRCA1 to sites of DNA DSBs (Wang & Elledge 2007; Wang et al. 2007). BRCA1 is also a ubiquitin ligase with both checkpoint and DNA repair functions, but its physiological substrates are unclear (Huen et al. 2010; Greenberg 2008). At its C-terminus, BRCA1 contains a pair of BRCT domains; these domains interact with Sae2/Ctp1/CtIP, suggesting that BRCA1 may regulate the nuclease activity of the MRN complex (Varma et al. 2005). The function of BRCA1 is of great interest, because the gene encoding BRCA1 is mutated in a subset of hereditary breast cancers (Xu & Solomon 1996).

53BP1 is also recruited to sites of DNA DSBs (Schultz et al. 2000). Its recruitment is mediated by a tudor domain that interacts with methylated lysines in histones H3 and H4 (K79 of histone H3 and K20 of histone H4) (Huyen et al. 2004; Botuyan et al. 2006; Schotta et al. 2008). Both these lysines are located in the nucleosome core, which suggests that recruitment of 53BP1 to sites of DNA DSBs must be accompanied by unstacking of the nucleosomes in the chromatin fiber. 53BP1 contains at its C-terminus a pair of BRCT domains, which are expected to interact with some phosphorylated protein, whose nature is still elusive (Zgheib et al. 2005). Interestingly, recruitment of 53BP1 to sites of DNA DSBs is dependent on H2AX phosphorylation and on MDC1 and RNF8 recruitment. However, how MDC1 and RNF8 facilitate recruitment of 53BP1 to sites of DNA DSBs is only partially resolved. As discussed above, H2AX phosphorylation recruits MDC1 to sites of DNA DSBs. MDC1 then recruits RNF8, leading to histone H2A and H2AX ubiquitylation. However, 53BP1 does not contain a ubiquitin-binding domain (Zgheib et al. 2009), so whether histone ubiquitylation facilitates 53BP1 recruitment to DNA damage sites remains unclear.

The Chk2 kinase is a downstream target of ATM (Jackson & Bartek 2009). Chk2 kinase activity is enhanced, when it becomes phosphorylated by ATM (Matsuoka et al. 2000; Melchionna et al. 2000). Both ATM and Chk2 phosphorylate the p53 tumor suppressor protein at serines 15 and 20, respectively (Canman et al. 1998; Chehab et al. 2000). p53 is a transcription factor with sequence-specific DNA binding activity (Vogelstein et al. 2000). The doubly-phosphorylated p53 protein becomes active and induces transient cell cycle arrest, permanent cell cycle arrest (senescence) or apoptosis (Clarke et al. 1993; Lowe et al. 1993; Di Leonardo et al. 1994; Canman et al. 1998; Chehab et al. 2000).

3.3 The Response to DNA Replication Stress

The term DNA replication stress is currently used to describe DNA replication fork stalling and/or collapse (Osborn et al. 2002; Halazonetis et al. 2008). The response of cells to DNA replication stress is of significant importance, because a defective response may introduce mutations and/or deletions in the genome (Fig. 3.2).

One of the first events after a DNA replication fork encounters a lesion that cannot be replicated is uncoupling of the DNA helicase from the DNA polymerase, creating a long stretch of single-stranded DNA that becomes coated by Replication Protein A (RPA). This is followed by recruitment of the checkpoint kinase ATR and stabilization of the fork (Zou & Elledge 2003; Cobb et al. 2005; Segurado & Diffley 2008). Then, either the fork backtracks (fork reversal) to expose the DNA lesion or a translesion polymerase is recruited to continue replication through the damaged DNA template. In the case of backtracking, the lesion may be repaired by DNA damage repair proteins or the nascent strands may be used as templates for DNA synthesis past the lesion via a process called template switching. If the stalled replication fork is not stabilized and the DNA replication machinery comes off the DNA (fork collapse), then the homologous recombination (HR) machinery will generate a DNA structure with a free 3′ end that can serve as a primer for DNA replication restart. Because this volume is focused on signaling, we will not discuss the proteins

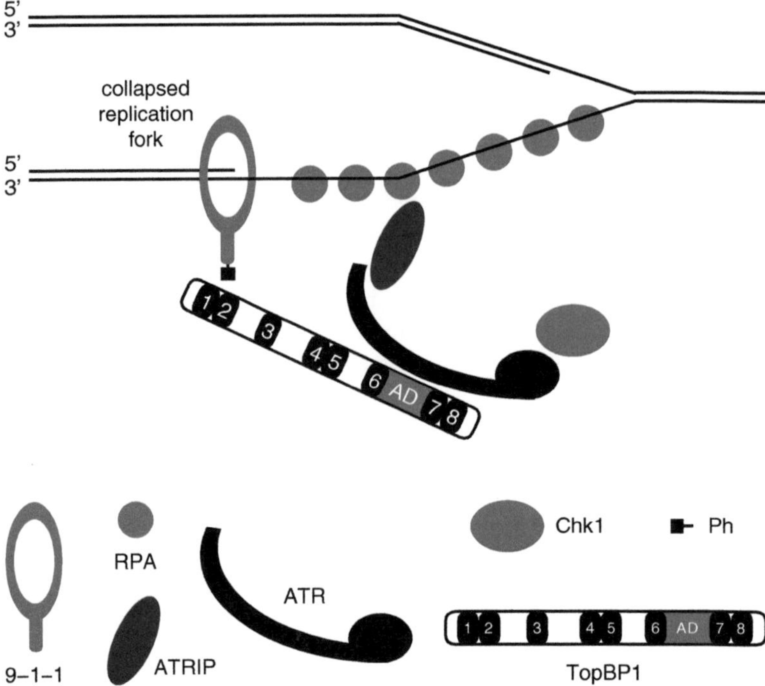

Fig. 3.2 DNA replications stress signaling pathways. A stalled or collapsed DNA replication fork leads to long stretches of single-stranded DNA, which, in turn, lead to the recruitment of RPA and ATRIP-ATR. The Rad9-Rad1-Hus1 (9-1-1) complex is also recruited at the junctions of single and double-stranded DNA. TopBP1 binds to phosphorylated Rad9 via its BRCT domains 1 and 2. The activation domain (AD) of TopBP1 facilitates activation of ATR, which, in turn, phosphorylates and activates Chk1. See text for more details. *Ph* phosphorylation

involved in the repair pathways that allow resumption of DNA replication after fork stalling or collapse; this topic has been covered recently by some excellent reviews (Friedel et al. 2009; Branzei & Foiani 2010).

Key proteins involved in the DNA replication checkpoint are the kinases ATM and Rad3-related (ATR) and Chk1; the ATR-interacting protein ATRIP; the Rad9-Rad1-Hus1 complex; and the proteins TopBP1, claspin, and BRCA1.

ATR and ATRIP form a stable protein complex (Cortez et al. 2001). ATR is a large protein kinase with high sequence similarity to ATM (Keegan et al. 1996; Tibbetts et al. 1999). ATR forms a constitutive protein complex with ATRIP. Together these two proteins are recruited to sites of DNA replication stress. This recruitment is mediated by ATRIP, which binds to the RPA that is coating the long stretches of single-stranded DNA that arise when DNA replication forks stall or collapse (Zou & Elledge 2003). The mechanism of ATR activation is unclear, but involves the Rad9-Rad1-Hus1 complex and the protein TopBP1 (Kumagai et al. 2006; Mordes et al. 2008; Navadgi-Patil & Burgers 2009).

The Rad9-Rad1-Hus1 (911) complex assembles as a hetero-trimeric ring that resembles structurally the PCNA ring (Dore et al. 2009). Like PCNA, the 911 ring is loaded on DNA by a clamp loader, consisting of RFC proteins. The PCNA and 911 clamp loaders are identical, except that one subunit of the PCNA clamp loader, RFC1, is replaced by Rad17 (Bermudez et al. 2003). Rad17, a substrate of ATR, recruits the 911 clamp loader to junctions of single and double-stranded DNA, that is to sites where DNA replication has ceased (Bermudez et al. 2003; Bao et al. 2001). The Rad9 protein of the 911 complex has an extended C-terminus, which is phosphorylated and recruits the protein TopBP1 to sites of DNA replication stress (Delacroix et al. 2007).

TopBP1 is a protein containing eight BRCT domains (Garcia et al. 2005). These domains apparently function to recruit TopBP1 to sites of DNA replication stress. Specifically, the pair of domains 1 and 2 interacts with the phosphorylated C-terminus of Rad9 and this interaction is required for recruitment of TopBP1 to sites of DNA replication stress (Delacroix et al. 2007). Between BRCT domains 6 and 7, TopBP1 contains a domain that interacts with ATR and upregulates its kinase activity (Kumagai et al. 2006). Thus, a model emerging for activation of ATR in response to DNA replication stress is that long stretches of single-stranded DNA recruit ATR-ATRIP and the Rad17 clamp loader. The latter loads the 911 complex, which in turn recruits TopBP1. Then TopBP1 either on its own or with the help of the 911 complex activates ATR.

Claspin is a protein that appears to be constitutively associated with the DNA replication fork (Kumagai & Dunphy 2000). Upon fork stalling, claspin recruits the Chk1 kinase to the stalled fork. This then facilitates phosphorylation and activation of Chk1 by ATR (Liu et al. 2006). Once activated, Chk1 phosphorylates and inactivates the Cdc25C phosphatase leading to inhibition of mitotic entry (Jackson & Bartek 2009; Osborn et al. 2002).

The BRCA1 protein, mentioned above in the context of DNA DSB signaling, is also recruited to sites of DNA replication stress and, like claspin, facilitates Chk1 activation (Huen et al. 2010; Greenberg 2008; Varma et al. 2005).

Finally, the tumor suppressor protein p53, mentioned above in the context of DNA DSB signaling, also participates in the response of cells to DNA replication stress. p53 becomes phosphorylated by the checkpoint kinases ATR and Chk1 (Tibbetts et al. 1999; Shieh et al. 2000). These phosphorylations enhance the activity of p53, leading to cell cycle arrest, apoptosis, and/or senescence depending on the magnitude of the DNA replication stress response and other factors, including cell type, cell cycle phase, etc.

3.4 DNA Replication Stress and DNA DSBs in Human Cancer

Several of the genes described above that participate in the response of cells to DNA DSBs or DNA replication stress are mutated in human cancer (Fig. 3.3). The most notable example is the p53 tumor suppressor gene, which is mutated in about half of

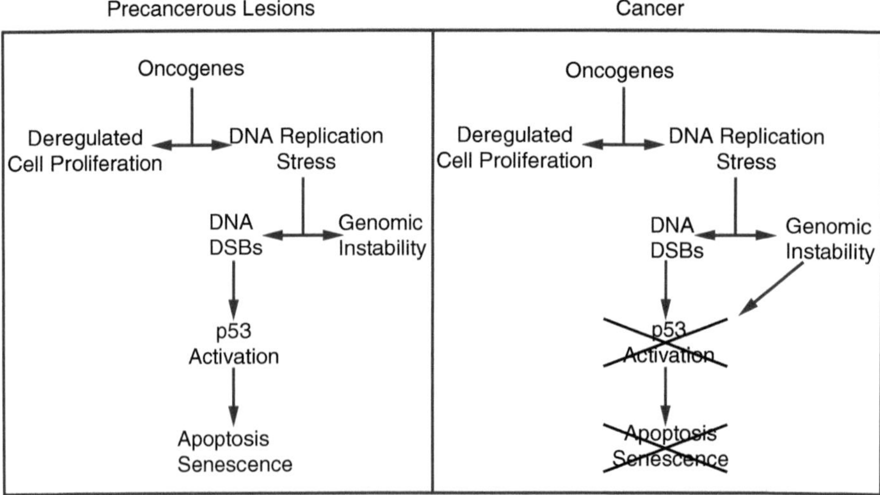

Fig. 3.3 Oncogene-induced DNA replication stress model for cancer development and progression. According to this model, activation of oncogenes is the first event that occurs in sporadic cancers. Oncogene activation leads to deregulated cell proliferation and DNA replication stress. The latter leads to DNA double-strand breaks (DSBs), which contribute to genomic instability and also activate p53, leading to apoptosis and/or senescence (precancerous lesions). In cancers, mutations that inactivate p53, allow the cells to proliferate without undergoing apoptosis or senescence

all human cancers (Hollstein et al. 1991; Sjoblom et al. 2006; Wood et al. 2007; Jones et al. 2008; Parsons et al. 2008; Cancer Genome Atlas Research Network 2008; Ding et al. 2008). Further, in other cancers that retain the wild-type p53 gene, p53 function is compromised by other mechanisms. These mechanisms include amplification of the genes encoding the MDM2 or MDM4 ubiquitin ligases, which target the p53 protein for degradation (Ding et al. 2008; Momand et al. 1992; Oliner et al. 1992), as well as expression of viral proteins, most notably the human papillomavirus E6 protein, which also targets p53 for degradation (Scheffner et al. 1993).

Several mechanisms have been proposed to explain the high frequency of p53 mutations in human cancer. One attractive mechanism postulates that inactivation of p53 promotes genomic instability, which thus facilitates the acquisition of oncogenic mutations. According to this model, p53 acts as a guardian of the genome (Lane 1992). However, p53 gene knockout experiments in both tissue culture cells and in mice have failed to demonstrate a significant increase in genomic instability after p53 inactivation (Kang et al. 2005; Bunz et al. 2002). Further, in precancerous lesions, genomic instability is evident before p53 mutations are acquired (Bartkova et al. 2005; Gorgoulis et al. 2005).

Another mechanism to explain the high frequency of p53 mutations invokes the presence of oncogenic stress in human cancers. The precise nature of this oncogenic stress was not defined, but this stress was proposed to lead to expression of the Alternate Reading Frame (ARF) tumor suppressor protein (Quelle et al. 1995; Zindy et al. 1998; de Stanchina et al. 1998). ARF would then lead to inhibition of MDM2 activity, increased levels of p53 and p53-dependent apoptosis (Zhang et al. 1998). Hence, developing cancer cells would experience a strong selective pressure favoring loss of function mutations in p53. Certain recent observations suggest that this mechanism appears to be more relevant for development of mouse cancers than for development of human cancers (see below).

A third hypothesis to explain the high frequency of p53 mutations in human cancers proposes that oncogenes induce DNA replication stress, which then leads to activation of p53. According to this mechanism, activated oncogenes drive cells to attempt DNA replication under inappropriate conditions, leading to persistent DNA replication stress (Halazonetis et al. 2008). In support of this claim, inhibitors of ATM and ATR suppress oncogene-induced apoptosis and senescence (Di Micco et al. 2006; Bartkova et al. 2006). Further, human precancerous lesions have signs of DNA DSBs and DNA replication stress before p53 mutations have been acquired (Bartkova et al. 2005; Gorgoulis et al. 2005). By immunohistochemistry, one observes the presence of phosphorylated histone H2AX, ATM, and Chk2, all indicating a DNA damage response in these early lesions. In addition, 53BP1 localizes to foci, suggestive of DNA DSBs, and p53 protein levels are high. In these precancerous lesions that retain wild-type p53 genes, the high levels of p53 protein lead to apoptosis or senescence and both these responses prevent rapid growth of the precancerous lesion. Essentially, p53 is acting as a tumor suppressor protein. At later stages in cancer development, the p53 genes become mutated or p53 function is otherwise inhibited. Such lesions escape p53-dependent apoptosis and senescence and are, therefore, clinically much more aggressive.

Several additional observations support the oncogene-induced DNA replication stress model for cancer development. The first relates to the presence of DNA replication stress in precancerous lesions and cancers. It has been established that DNA replication stress, a term referring to collapse/disassembly of DNA replication forks, is often accompanied by deletions of specific sites in the genome called common fragile sites (Arlt et al. 2006; Casper et al. 2002). Apparently, these genomic sites are hard to replicate and when DNA replication is compromised, cells divide with unreplicated DNA, leading to deletions that target these sites. The genomic instability in human precancerous lesions targets preferentially the common fragile sites, arguing that the presence of DNA damage in these lesions is due to DNA replication stress (Bartkova et al. 2005; Gorgoulis et al. 2005; Di Micco et al. 2006).

In experimental settings, oncogene activation also leads to DNA replication stress. This is evident by the presence of nuclear foci containing (RPA) in cells overexpressing oncogenes, as well as by the presence of prematurely-terminated replication forks, whose presence can be detected by the DNA combing method (Di Micco et al. 2006; Bartkova et al. 2006).

The spectrum of mutations in human cancers further supports the oncogene-induced DNA replication stress model. Consistent with this model, mutations in cancer target not only p53, but also other genes that function in the DNA damage response pathway. The most well-known example is mutations targeting ATM (Ding et al. 2008; Greenman et al. 2007). Further, consistent with the model that p53 in cancer is activated by DNA replication stress and DNA DSBs, p53 and ATM mutations were mutually exclusive in a sample of 188 lung adenocarcinomas (Ding et al. 2008; Negrini et al. 2010). This implies that in those tumors in which the ATM gene is mutated, there is no need to inactivate p53, because ATM is the main signaling kinase leading to p53 activation in cells with DNA damage. In contrast, mutations targeting ARF are not mutually exclusive with p53 mutations, suggesting that ARF is not the principal agent leading to p53 activation in human cancers (Negrini et al. 2010).

3.5 Future Directions

DNA damage and DNA replication stress signaling pathways are clearly implicated in cancer development, as described above. In turn, this implies that a better understanding of these pathways at the molecular level and in actual human cancers could lead to the identification of targets for novel cancer therapies. We anticipate that in the next few years, high-throughput siRNA screens will identify such targets. Thus, new opportunities for cancer therapies may be just around the corner.

References

Adelman CA, Petrini JH (2009) Division of labor: DNA repair and the cell cycle specific functions of the Mre11 complex. Cell Cycle 8:1510–1514

Arlt MF, Durkin SG, Ragland RL, Glover TW (2006) Common fragile sites as targets for chromosome rearrangements. DNA Repair 5:1126–1135

Bakkenist CJ, Kastan MB (2003) DNA damage activates ATM through intermolecular autophosphorylation and dimer dissociation. Nature 421:499–506

Bao S, Tibbetts RS, Brumbaugh KM, Fang Y, Richardson DA, Ali A, Chen SM, Abraham RT, Wang XF (2001) ATR/ATM-mediated phosphorylation of human Rad17 is required for genotoxic stress responses. Nature 411:969–974

Bartkova J, Horejsi Z, Koed K, Kramer A, Tort F, Zieger K, Guldberg P, Sehested M, Nesland JM, Lukas C, Orntoft T, Lukas J, Bartek J (2005) DNA damage response as a candidate anti-cancer barrier in early human tumorigenesis. Nature 434:864–870

Bartkova J, Rezaei N, Liontos M, Karakaidos P, Kletsas D, Issaeva N, Vassiliou LV, Kolettas E, Niforou K, Zoumpourlis VC, Takaoka M, Nakagawa H, Tort F, Fugger K, Johansson F, Sehested M, Andersen CL, Dyrskjot L, Orntoft T, Lukas J, Kittas C, Helleday T, Halazonetis TD, Bartek J, Gorgoulis VG (2006) Oncogene-induced senescence is part of the tumorigenesis barrier imposed by DNA damage checkpoints. Nature 444:633–637

Bermudez VP, Lindsey-Boltz LA, Cesare AJ, Maniwa Y, Griffith JD, Hurwitz J, Sancar A (2003) Loading of the human 9-1-1 checkpoint complex onto DNA by the checkpoint clamp loader hRad17-replication factor C complex in vitro. Proc Natl Acad Sci U S A 100:1633–1638

Bishop JM (1991) Molecular themes in oncogenesis. Cell 64:235–248

Botuyan MV, Lee J, Ward IM, Kim JE, Thompson JR, Chen J, Mer G (2006) Structural basis for the methylation state-specific recognition of histone H4-K20 by 53BP1 and Crb2 in DNA repair. Cell 127:1361–1373

Branzei D, Foiani M (2010) Maintaining genome stability at the replication fork. Nat Rev Mol Cell Biol 11:208–219

Bunz F, Fauth C, Speicher MR, Dutriaux A, Sedivy JM, Kinzler KW, Vogelstein B, Lengauer C (2002) Targeted inactivation of p53 in human cells does not result in aneuploidy. Cancer Res 62:1129–1133

Cancer Genome Atlas Research Network (2008) Comprehensive genomic characterization defines human glioblastoma genes and core pathways. Nature 455:1061–1068

Canman CE, Lim DS, Cimprich KA, Taya Y, Tamai K, Sakaguchi K, Appella E, Kastan MB, Siliciano JD (1998) Activation of the ATM kinase by ionizing radiation and phosphorylation of p53. Science 281:1677–1679

Casper AM, Nghiem P, Arlt MF, Glover TW (2002) ATR regulates fragile site stability. Cell 111:779–789

Chehab NH, Malikzay A, Appel M, Halazonetis TD (2000) Chk2/hCds1 functions as a DNA damage checkpoint in G(1) by stabilizing p53. Genes Dev 14:278–288

Clarke AR, Purdie CA, Harrison DJ, Morris RG, Bird CC, Hooper ML, Wyllie AH (1993) Thymocyte apoptosis induced by p53-dependent and independent pathways. Nature 362:849–852

Cobb JA, Schleker T, Rojas V, Bjergbaek L, Tercero JA, Gasser SM (2005) Replisome instability, fork collapse, and gross chromosomal rearrangements arise synergistically from Mec1 kinase and RecQ helicase mutations. Genes Dev 19:3055–3069

Cortez D, Guntuku S, Qin J, Elledge SJ (2001) ATR and ATRIP: partners in checkpoint signaling. Science 294:1713–1716

Daniel JA, Pellegrini M, Lee JH, Paull TT, Feigenbaum L, Nussenzweig A (2008) Multiple autophosphorylation sites are dispensable for murine ATM activation in vivo. J Cell Biol 183:777–783

de Stanchina E, McCurrach ME, Zindy F, Shieh SY, Ferbeyre G, Samuelson AV, Prives C, Roussel MF, Sherr CJ, Lowe SW (1998) E1A signaling to p53 involves the p19(ARF) tumor suppressor. Genes Dev 12:2434–2442

Delacroix S, Wagner JM, Kobayashi M, Yamamoto K, Karnitz LM (2007) The Rad9-Hus1-Rad1 (9-1-1) clamp activates checkpoint signaling via TopBP1. Genes Dev 21:1472–1477

Di Leonardo A, Linke SP, Clarkin K, Wahl GM (1994) DNA damage triggers a prolonged p53-dependent G1 arrest and long-term induction of Cip1 in normal human fibroblasts. Genes Dev 8:2540–2551

Di Micco R, Fumagalli M, Cicalese A, Piccinin S, Gasparini P, Luise C, Schurra C, Garre' M, Nuciforo PG, Bensimon A, Maestro R, Pelicci PG, d'Adda di Fagagna F (2006) Oncogene-induced senescence is a DNA damage response triggered by DNA hyper-replication. Nature 444:638–642

Ding L, Getz G, Wheeler DA, Mardis ER, McLellan MD, Cibulskis K, Sougnez C, Greulich H, Muzny DM, Morgan MB, Fulton L, Fulton RS, Zhang Q, Wendl MC, Lawrence MS, Larson DE, Chen K, Dooling DJ, Sabo A, Hawes AC, Shen H, Jhangiani SN, Lewis LR, Hall O, Zhu Y, Mathew T, Ren Y, Yao J, Scherer SE, Clerc K, Metcalf GA, Ng B, Milosavljevic A, Gonzalez-Garay ML, Osborne JR, Meyer R, Shi X, Tang Y, Koboldt DC, Lin L, Abbott R, Miner TL, Pohl C, Fewell G, Haipek C, Schmidt H, Dunford-Shore BH, Kraja A, Crosby SD, Sawyer CS, Vickery T, Sander S, Robinson J, Winckler W, Baldwin J, Chirieac LR, Dutt A, Fennell T, Hanna M, Johnson BE, Onofrio RC, Thomas RK, Tonon G, Weir BA, Zhao X, Ziaugra L, Zody MC, Giordano T, Orringer MB, Roth JA, Spitz MR, Wistuba II, Ozenberger B, Good PJ, Chang AC, Beer DG, Watson MA, Ladanyi M, Broderick S, Yoshizawa A, Travis WD, Pao W, Province MA, Weinstock GM, Varmus HE, Gabriel SB, Lander ES, Gibbs RA, Meyerson M, Wilson RK (2008) Somatic mutations affect key pathways in lung adenocarcinoma. Nature 455:1069–1075

Doil C, Mailand N, Bekker-Jensen S, Menard P, Larsen DH, Pepperkok R, Ellenberg J, Panier S, Durocher D, Bartek J, Lukas J, Lukas C (2009) RNF168 binds and amplifies ubiquitin conjugates on damaged chromosomes to allow accumulation of repair proteins. Cell 136:435–446

Dore AS, Kilkenny ML, Rzechorzek NJ, Pearl LH (2009) Crystal structure of the rad9-rad1-hus1 DNA damage checkpoint complex–implications for clamp loading and regulation. Mol Cell 34:735–745

Falck J, Coates J, Jackson SP (2005) Conserved modes of recruitment of ATM, ATR and DNA-PKcs to sites of DNA damage. Nature 434:605–611

Friedberg EC (1991) Yeast genes involved in DNA-repair processes: new looks on old faces. Mol Microbiol 5:2303–2310

Friedel AM, Pike BL, Gasser SM (2009) ATR/Mec1: coordinating fork stability and repair. Curr Opin Cell Biol 21:237–244

Garcia V, Furuya K, Carr AM (2005) Identification and functional analysis of TopBP1 and its homologs. DNA Repair 4:1227–1239

Goldberg M, Stucki M, Falck J, D'Amours D, Rahman D, Pappin D, Bartek J, Jackson SP (2003) MDC1 is required for the intra-S-phase DNA damage checkpoint. Nature 421:952–956

Gorgoulis VG, Vassiliou LV, Karakaidos P, Zacharatos P, Kotsinas A, Liloglou T, Venere M, Ditullio RA Jr, Kastrinakis NG, Levy B, Kletsas D, Yoneta A, Herlyn M, Kittas C, Halazonetis TD (2005) Activation of the DNA damage checkpoint and genomic instability in human precancerous lesions. Nature 434:907–913

Greenberg RA (2008) Recognition of DNA double strand breaks by the BRCA1 tumor suppressor network. Chromosoma 117:305–317

Greenman C, Stephens P, Smith R, Dalgliesh GL, Hunter C, Bignell G, Davies H, Teague J, Butler A, Stevens C, Edkins S, O'Meara S, Vastrik I, Schmidt EE, Avis T, Barthorpe S, Bhamra G, Buck G, Choudhury B, Clements J, Cole J, Dicks E, Forbes S, Gray K, Halliday K, Harrison R, Hills K, Hinton J, Jenkinson A, Jones D, Menzies A, Mironenko T, Perry J, Raine K, Richardson D, Shepherd R, Small A, Tofts C, Varian J, Webb T, West S, Widaa S, Yates A, Cahill DP, Louis DN, Goldstraw P, Nicholson AG, Brasseur F, Looijenga L, Weber BL, Chiew YE, DeFazio A, Greaves MF, Green AR, Campbell P, Birney E, Easton DF, Chenevix-Trench G, Tan MH, Khoo SK, Teh BT, Yuen ST, Leung SY, Wooster R, Futreal PA, Stratton MR (2007) Patterns of somatic mutation in human cancer genomes. Nature 446:153–158

Halazonetis TD, Gorgoulis VG, Bartek J (2008) An oncogene-induced DNA damage model for cancer development. Science 319:1352–1355

Hartwell LH, Weinert TA (1989) Checkpoints: controls that ensure the order of cell cycle events. Science 246:629–634

Hollstein M, Sidransky D, Vogelstein B, Harris CC (1991) p53 mutations in human cancers. Science 253:49–53

Hopfner KP, Karcher A, Shin DS, Craig L, Arthur LM, Carney JP, Tainer JA (2000) Structural biology of Rad50 ATPase: ATP-driven conformational control in DNA double-strand break repair and the ABC-ATPase superfamily. Cell 101:789–800

Hopfner KP, Karcher A, Craig L, Woo TT, Carney JP, Tainer JA (2001) Structural biochemistry and interaction architecture of the DNA double-strand break repair Mre11 nuclease and Rad50-ATPase. Cell 105:473–485

Hopfner KP, Craig L, Moncalian G, Zinkel RA, Usui T, Owen BA, Karcher A, Henderson B, Bodmer JL, McMurray CT, Carney JP, Petrini JH, Tainer JA (2002) The Rad50 zinc-hook is a structure joining Mre11 complexes in DNA recombination and repair. Nature 418:562–566

Huen MS, Grant R, Manke I, Minn K, Yu X, Yaffe MB, Chen J (2007) RNF8 transduces the DNA-damage signal via histone ubiquitylation and checkpoint protein assembly. Cell 131:901–914

Huen MS, Sy SM, Chen J (2010) BRCA1 and its toolbox for the maintenance of genome integrity. Nat Rev Mol Cell Biol 11:138–148

Huyen Y, Zgheib O, Ditullio RA Jr, Gorgoulis VG, Zacharatos P, Petty TJ, Sheston EA, Mellert HS, Stavridi ES, Halazonetis TD (2004) Methylated lysine 79 of histone H3 targets 53BP1 to DNA double-strand breaks. Nature 432:406–411

Ivanov EL, Haber JE (1997) DNA repair: RAD alert. Curr Biol 7:R492–R495

Jackson SP, Bartek J (2009) The DNA-damage response in human biology and disease. Nature 461:1071–1078

Jones S, Zhang X, Parsons DW, Lin JC, Leary RJ, Angenendt P, Mankoo P, Carter H, Kamiyama H, Jimeno A, Hong SM, Fu B, Lin MT, Calhoun ES, Kamiyama M, Walter K, Nikolskaya T, Nikolsky Y, Hartigan J, Smith DR, Hidalgo M, Leach SD, Klein AP, Jaffee EM, Goggins M, Maitra A, Iacobuzio-Donahue C, Eshleman JR, Kern SE, Hruban RH, Karchin R, Papadopoulos N, Parmigiani G, Vogelstein B, Velculescu VE, Kinzler KW (2008) Core signaling pathways in human pancreatic cancers revealed by global genomic analyses. Science 321:1801–1806

Kang J, Ferguson D, Song H, Bassing C, Eckersdorff M, Alt FW, Xu Y (2005) Functional interaction of H2AX, NBS1, and p53 in ATM-dependent DNA damage responses and tumor suppression. Mol Cell Biol 25:661–670

Keegan KS, Holtzman DA, Plug AW, Christenson ER, Brainerd EE, Flaggs G, Bentley NJ, Taylor EM, Meyn MS, Moss SB, Carr AM, Ashley T, Hoekstra MF (1996) The Atr and Atm protein kinases associate with different sites along meiotically pairing chromosomes. Genes Dev 10:2423–2437

Keith CT, Schreiber SL (1995) PIK-related kinases: DNA repair, recombination, and cell cycle checkpoints. Science 270:50–51

Kinzler KW, Vogelstein B (1997) Cancer-susceptibility genes. Gatekeepers and caretakers. Nature 386:761–763

Klein G (1987) The approaching era of the tumor suppressor genes. Science 238:1539–1545

Kumagai A, Dunphy WG (2000) Claspin, a novel protein required for the activation of Chk1 during a DNA replication checkpoint response in Xenopus egg extracts. Mol Cell 6:839–849

Kumagai A, Lee J, Yoo HY, Dunphy WG (2006) TopBP1 activates the ATR-ATRIP complex. Cell 124:943–955

Lane DP (1992) Cancer. p53, guardian of the genome. Nature 358:15–16

Lee JH, Paull TT (2005) ATM activation by DNA double-strand breaks through the Mre11-Rad50-Nbs1 complex. Science 308:551–554

Levine AJ (1990) Tumor suppressor genes. Bioessays 12:60–66

Liu S, Bekker-Jensen S, Mailand N, Lukas C, Bartek J, Lukas J (2006) Claspin operates downstream of TopBP1 to direct ATR signaling towards Chk1 activation. Mol Cell Biol 26:6056–6064

Lloyd J, Chapman JR, Clapperton JA, Haire LF, Hartsuiker E, Li J, Carr AM, Jackson SP, Smerdon SJ (2009) A supramodular FHA/BRCT-repeat architecture mediates Nbs1 adaptor function in response to DNA damage. Cell 139:100–111

Loeb LA (1991) Mutator phenotype may be required for multistage carcinogenesis. Cancer Res 51:3075–3079

Lowe SW, Schmitt EM, Smith SW, Osborne BA, Jacks T (1993) p53 is required for radiation-induced apoptosis in mouse thymocytes. Nature 362:847–849

Mailand N, Bekker-Jensen S, Faustrup H, Melander F, Bartek J, Lukas C, Lukas J (2007) RNF8 ubiquitylates histones at DNA double-strand breaks and promotes assembly of repair proteins. Cell 131:887–900

Manning G, Whyte DB, Martinez R, Hunter T, Sudarsanam S (2002) The protein kinase complement of the human genome. Science 298:1912–1934

Matsuoka S, Rotman G, Ogawa A, Shiloh Y, Tamai K, Elledge SJ (2000) Ataxia telangiectasia-mutated phosphorylates Chk2 in vivo and in vitro. Proc Natl Acad Sci U S A 97:10389–10394

Melchionna R, Chen XB, Blasina A, McGowan CH (2000) Threonine 68 is required for radiation-induced phosphorylation and activation of Cds1. Nat Cell Biol 2:762–765

Momand J, Zambetti GP, Olson DC, George D, Levine AJ (1992) The mdm-2 oncogene product forms a complex with the p53 protein and inhibits p53-mediated transactivation. Cell 69:1237–1245

Morales M, Theunissen JW, Kim CF, Kitagawa R, Kastan MB, Petrini JH (2005) The Rad50S allele promotes ATM-dependent DNA damage responses and suppresses ATM deficiency: implications for the Mre11 complex as a DNA damage sensor. Genes Dev 19:3043–3054

Mordes DA, Glick GG, Zhao R, Cortez D (2008) TopBP1 activates ATR through ATRIP and a PIKK regulatory domain. Genes Dev 22:1478–1489

Navadgi-Patil VM, Burgers PM (2009) The unstructured C-terminal tail of the 9-1-1 clamp subunit Ddc1 activates Mec1/ATR via two distinct mechanisms. Mol Cell 36:743–753

Negrini S, Gorgoulis VG, Halazonetis TD (2010) Genomic instability – an evolving hallmark of cancer. Nat Rev Mol Cell Biol 11:220–228

Oliner JD, Kinzler KW, Meltzer PS, George DL, Vogelstein B (1992) Amplification of a gene encoding a p53-associated protein in human sarcomas. Nature 358:80–83

Osborn AJ, Elledge SJ, Zou L (2002) Checking on the fork: the DNA-replication stress-response pathway. Trends Cell Biol 12:509–516

Parada LF, Tabin CJ, Shih C, Weinberg RA (1982) Human EJ bladder carcinoma oncogene is homologue of Harvey sarcoma virus ras gene. Nature 297:474–478

Parsons DW, Jones S, Zhang X, Lin JC, Leary RJ, Angenendt P, Mankoo P, Carter H, Siu IM, Gallia GL, Olivi A, McLendon R, Rasheed BA, Keir S, Nikolskaya T, Nikolsky Y, Busam DA, Tekleab H, Diaz LA Jr, Hartigan J, Smith DR, Strausberg RL, Marie SK, Shinjo SM, Yan H, Riggins GJ, Bigner DD, Karchin R, Papadopoulos N, Parmigiani G, Vogelstein B, Velculescu VE, Kinzler KW (2008) An integrated genomic analysis of human glioblastoma multiforme. Science 321:1807–1812

Paulovich AG, Toczyski DP, Hartwell LH (1997) When checkpoints fail. Cell 88:315–321

Perry J, Kleckner N (2003) The ATRs, ATMs, and TORs are giant HEAT repeat proteins. Cell 112:151–155

Quelle DE, Zindy F, Ashmun RA, Sherr CJ (1995) Alternative reading frames of the INK4a tumor suppressor gene encode two unrelated proteins capable of inducing cell cycle arrest. Cell 83:993–1000

Redon C, Pilch D, Rogakou E, Sedelnikova O, Newrock K, Bonner W (2002) Histone H2A variants H2AX and H2AZ. Curr Opin Genet Dev 12:162–169

Rogakou EP, Pilch DR, Orr AH, Ivanova VS, Bonner WM (1998) DNA double-stranded breaks induce histone H2AX phosphorylation on serine 139. J Biol Chem 273:5858–5868

Rogakou EP, Boon C, Redon C, Bonner WM (1999) Megabase chromatin domains involved in DNA double-strand breaks in vivo. J Cell Biol 146:905–916

Sartori AA, Lukas C, Coates J, Mistrik M, Fu S, Bartek J, Baer R, Lukas J, Jackson SP (2007) Human CtIP promotes DNA end resection. Nature 450:509–514

Scheffner M, Huibregtse JM, Vierstra RD, Howley PM (1993) The HPV-16 E6 and E6-AP complex functions as a ubiquitin-protein ligase in the ubiquitination of p53. Cell 75:495–505

Schotta G, Sengupta R, Kubicek S, Malin S, Kauer M, Callèn E, Celeste A, Pagani M, Opravil S, De La Rosa-Velazquez IA, Espejo A, Bedford MT, Nussenzweig A, Busslinger M, Jenuwein T (2008) A chromatin-wide transition to H4K20 monomethylation impairs genome integrity and programmed DNA rearrangements in the mouse. Genes Dev 22:2048–2061

Schultz LB, Chehab NH, Malikzay A, Halazonetis TD (2000) p53 binding protein 1 (53BP1) is an early participant in the cellular response to DNA double-strand breaks. J Cell Biol 151:1381–1390

Segurado M, Diffley JF (2008) Separate roles for the DNA damage checkpoint protein kinases in stabilizing DNA replication forks. Genes Dev 22:1816–1827

Shieh SY, Ahn J, Tamai K, Taya Y, Prives C (2000) The human homologs of checkpoint kinases Chk1 and Cds1 (Chk2) phosphorylate p53 at multiple DNA damage-inducible sites. Genes Dev 14:289–300

Shiloh Y (2003) ATM and related protein kinases: safeguarding genome integrity. Nat Rev Cancer 3:155–168

Sjoblom T, Jones S, Wood LD, Parsons DW, Lin J, Barber TD, Mandelker D, Leary RJ, Ptak J, Silliman N, Szabo S, Buckhaults P, Farrell C, Meeh P, Markowitz SD, Willis J, Dawson D, Willson JK, Gazdar AF, Hartigan J, Wu L, Liu C, Parmigiani G, Park BH, Bachman KE, Papadopoulos N, Vogelstein B, Kinzler KW, Velculescu VE (2006) The consensus coding sequences of human breast and colorectal cancers. Science 314:268–274

Stewart GS, Panier S, Townsend K, Al-Hakim AK, Kolas NK, Miller ES, Nakada S, Ylanko J, Olivarius S, Mendez M, Oldreive C, Wildenhain J, Tagliaferro A, Pelletier L, Taubenheim N, Durandy A, Byrd PJ, Stankovic T, Taylor AM, Durocher D (2009) The RIDDLE syndrome protein mediates a ubiquitin-dependent signaling cascade at sites of DNA damage. Cell 136:420–434

Stucki M, Clapperton JA, Mohammad D, Yaffe MB, Smerdon SJ, Jackson SP (2005) MDC1 directly binds phosphorylated histone H2AX to regulate cellular responses to DNA double-strand breaks. Cell 123:1213–1226

Tibbetts RS, Brumbaugh KM, Williams JM, Sarkaria JN, Cliby WA, Shieh SY, Taya Y, Prives C, Abraham RT (1999) A role for ATR in the DNA damage-induced phosphorylation of p53. Genes Dev 13:152–157

Varma AK, Brown RS, Birrane G, Ladias JA (2005) Structural basis for cell cycle checkpoint control by the BRCA1-CtIP complex. Biochemistry 44:10941–10946

Vogelstein B, Lane D, Levine AJ (2000) Surfing the p53 network. Nature 408:307–310

Wang B, Elledge SJ (2007) Ubc13/Rnf8 ubiquitin ligases control foci formation of the Rap80/Abraxas/Brca1/Brcc36 complex in response to DNA damage. Proc Natl Acad Sci U S A 104:20759–20763

Wang B, Matsuoka S, Ballif BA, Zhang D, Smogorzewska A, Gygi SP, Elledge SJ (2007) Abraxas and RAP80 form a BRCA1 protein complex required for the DNA damage response. Science 316:1194–1198

Williams RS, Williams JS, Tainer JA (2007) Mre11-Rad50-Nbs1 is a keystone complex connecting DNA repair machinery, double-strand break signaling, and the chromatin template. Biochem Cell Biol 85:509–520

Williams RS, Moncalian G, Williams JS, Yamada Y, Limbo O, Shin DS, Groocock LM, Cahill D, Hitomi C, Guenther G, Moiani D, Carney JP, Russell P, Tainer JA (2008) Mre11 dimers coordinate DNA end bridging and nuclease processing in double-strand-break repair. Cell 135:97–109

Williams RS, Dodson GE, Limbo O, Yamada Y, Williams JS, Guenther G, Classen S, Glover JN, Iwasaki H, Russell P, Tainer JA (2009) Nbs1 flexibly tethers Ctp1 and Mre11-Rad50 to coordinate DNA double-strand break processing and repair. Cell 139:87–99

Wood LD, Parsons DW, Jones S, Lin J, Sjöblom T, Leary RJ, Shen D, Boca SM, Barber T, Ptak J, Silliman N, Szabo S, Dezso Z, Ustyanksky V, Nikolskaya T, Nikolsky Y, Karchin R, Wilson PA, Kaminker JS, Zhang Z, Croshaw R, Willis J, Dawson D, Shipitsin M, Willson JK, Sukumar S, Polyak K, Park BH, Pethiyagoda CL, Pant PV, Ballinger DG, Sparks AB, Hartigan J, Smith DR, Suh E, Papadopoulos N, Buckhaults P, Markowitz SD, Parmigiani G, Kinzler KW, Velculescu VE, Vogelstein B (2007) The genomic landscapes of human breast and colorectal cancers. Science 318:1108–1113

Xu CF, Solomon E (1996) Mutations of the BRCA1 gene in human cancer. Semin Cancer Biol 7:33–40

Zgheib O, Huyen Y, DiTullio RA Jr, Snyder A, Venere M, Stavridi ES, Halazonetis TD (2005) ATM signaling and 53BP1. Radiother Oncol 76:119–122

Zgheib O, Pataky K, Brugger J, Halazonetis TD (2009) An oligomerized 53BP1 tudor domain suffices for recognition of DNA double-strand breaks. Mol Cell Biol 29:1050–1058

Zhang Y, Xiong Y, Yarbrough WG (1998) ARF promotes MDM2 degradation and stabilizes p53: ARF-INK4a locus deletion impairs both the Rb and p53 tumor suppression pathways. Cell 92:725–734

Zindy F, Eischen CM, Randle DH, Kamijo T, Cleveland JL, Sherr CJ, Roussel MF (1998) Myc signaling via the ARF tumor suppressor regulates p53-dependent apoptosis and immortalization. Genes Dev 12:2424–2433

Zou L, Elledge SJ (2003) Sensing DNA damage through ATRIP recognition of RPA-ssDNA complexes. Science 300:1542–1548

Chapter 4
Nonreceptor Tyrosine Kinases and Their Roles in Cancer

Jon R. Wiener and Gary E. Gallick

4.1 Historical Perspective

In 1911, Dr. Peyton Rous, a young American pathologist newly hired as the Director of the Laboratory for Cancer Research at the Rockefeller Institute in New York City, discovered that a transmissible agent caused a spindle-cell sarcoma that spontaneously arose in a Plymouth Rock hen. Utilizing Koch's postulates as a guideline, Rous was able to demonstrate that bacteria- and cell-free filtrate, or supernatants from tumor cells, contained a transmissible virus that was the etiologic agent of the chicken tumors (Rous 1911). What he discovered was the first oncogenic virus, now named Rous sarcoma virus (RSV), in his honor. In retrospect, this discovery was monumental, but for decades was met with indifference, as viruses that caused rare chicken tumors were assumed to be of little relevance to etiology of human tumors.

In the late 1950s, Temin and Rubin modified a standard focus-forming assay to show that a single RSV could transform tissue culture cells and generate a 'focus' of oncogenic transformation. This method was further utilized to isolate a number of additional viruses that could induce transformation in tissue culture and tumors in appropriate animal models, and opened the door to the isolation of tumor-forming *retro*viruses, known for their ability to direct DNA synthesis from an RNA genome through a polymerase commonly termed reverse transcriptase (or more appropriately, RNA-directed DNA polymerase). The double-stranded copy of the viral DNA then was able to integrate into the host genome, and expression of viral oncogenes was shown to be sufficient to induce malignant transformation (Martin 1970). The growing interest in studying replication of these viruses and mechanisms by which their viral oncogenes induced malignant transformation led to Rous winning the Nobel Prize in 1966 for his seminal work.

J.R. Wiener • G.E. Gallick (✉)
Anderson Cancer Center, Houston, TX 77030, USA
e-mail: ggallick@mdanderson.org

D.A. Frank (ed.), *Signaling Pathways in Cancer Pathogenesis and Therapy*,
DOI 10.1007/978-1-4614-1216-8_4, © Springer Science+Business Media, LLC 2012

The next monumental advance attributable to Src (nomenclature: the protein is denoted as Src, the gene as *src*: abbreviated from Sarcoma) was identification of the gene responsible for malignant transformation. RSV was shown to possess in its genome, in addition to the viral structural and polymerase genes, a gene known as v-*src* (v-designating viral origin, c- designating cellular origin), the protein product of which was oncogenic. Later, in the 1970s, Stehelin, Bishop, and Varmus showed that v-*src* was a mutated and constitutively activated form of a cellular homolog (proto-oncogene) known as c-*src*. They postulated that the virus transduced this gene, in part, due to the relative infidelity of reverse transcriptase, a discovery that led Bishop and Varmus to win the Nobel Prize in 1989 (Stehelin et al. 1976). Yet another path-breaking discovery attributable to Src was that it was the first protein shown to contain intrinsic protein tyrosine kinase (PTK) activity (Hunter & Sefton 1980).

Thus, the Src nonreceptor tyrosine kinase set the paradigm for understanding numerous processes in malignant transformation. However, as more and more altered proto-oncogenes were discovered in human tumors, interest in Src and its family of PTKs waned. Rather, interest focused on proto-oncogenes affected by chromosomal rearrangements, gene amplification, and mutation. Research in exploring the Src protein and its signaling functions as they relate to human cancers did not proceed rapidly, likely because c-*src* has very rarely been reported to be mutated, amplified, or rearranged in human cancers. More recently, however, investigators have recognized the importance of Src overexpression and/or aberrant activation in many human tumors (reviewed in Summy & Gallick 2003).

4.2   The SRC Family of Tyrosine Kinases

The Src family of PTKs (Src Family Kinases, SFK) is composed of nine members: Src, Yes, Lyn, Fyn, Lck, Hck, Blk, Fgr, and Yrk. Except for a unique domain near the amino terminus, all family members are structurally conserved with the kinase (Src Homology, SH-1) domain the most conserved domain in the family. All of the family members thus far studied are myristoylated at the amino terminus, and some are palmitoylated (see below), and thus primarily function by association with the inner leaflet of the cellular plasma membrane. They function in numerous cell signaling pathways, including the Ras/MAPK, p38, Stat 3, PI3K, and JNK, growth factor receptor and β-adrenoreceptors receptor pathways, among others, and thereby modulate proliferation, survival, invasiveness, motility, angiogenesis, stress response, and numerous other physiologic functions (Fig. 4.1).

SFKs are expressed in a multitude of cell lineages, which gives a clue as to their function. For example, Src, Fyn, Yes, and Lyn are found rather ubiquitously in multiple cell lineages, and coexpressed in some lineages, while Hck, Fgr, Blk, and Lck are primarily expressed in hematopoietic cell lineages. In all cells where they are expressed, SFKs interact with, are activated by, and modulate the function of many cytokine receptors, G-protein-coupled receptors (GPCR), growth factor receptor tyrosine kinases (GFRTK), and integrins (Thomas & Brugge 1997). That multiple

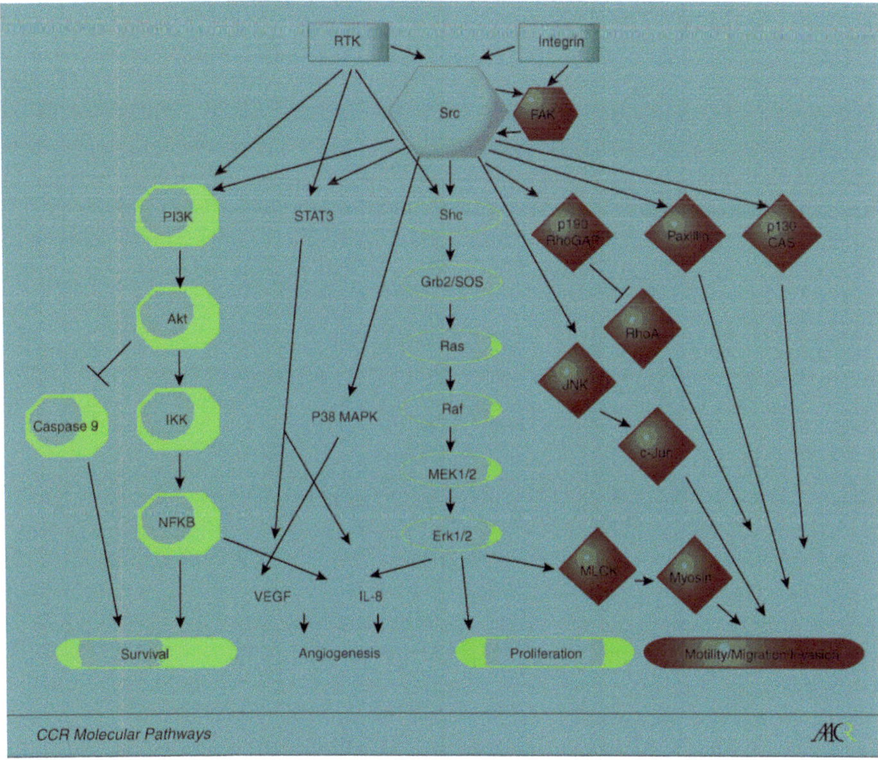

Fig. 4.1 Src and its family members play a central role in cellular signal transduciton

SFKs are expressed in some cells may explain the redundancy observed in some knockout models (see below).

As with most eukaryotic proteins, subcellular localization is critical to function and is thus tightly regulated. SFKs, by virtue of acylation at their amino terminus (see below), predominantly localize to cellular membranes, the plasma membrane being the major membrane associated with these proteins. The association of SFKs to the plasma membrane is essential for some Src functions, but it is by no means the only membrane in the cell associated with SFKs – they have also been found associated with endosomal membranes, rough endoplasmic reticulum, and secretory vesicles, and in the nucleus, but the function of SFKs in these latter locations remains unclear. When in association with the inner leaflet of the cellular plasma membrane, they are in close proximity to, and therefore interact with, the carboxy-terminal tails of cellular growth factor receptors, usually GFRTK or GPCR, and thereby are involved in the cellular signaling cascade that emanates from growth factor receptor PTK activation by ligands in normal cells, or, in the case of cancer cells, that can emanate from receptors in a ligand-independent manner (Thomas & Brugge 1997; Basu 2004). Very recently, Src has been shown to cotranslocate with

EGF-R to the mitochondria (Demory et al. 2009). Src thus occupies a central focus in a variety of signal transduction cascades vital for cellular functions.

Alterations in many signaling pathways lead to increased Src activation, and increased Src activation has been shown to contribute to the malignant phenotype not only in tumor cells but also in cells of the tumor microenvironment. For these reasons, multiple SFK inhibitors have been developed and are in clinical trial for advanced stage cancers of many organs. In this chapter, we briefly describe Src structure and function, the role of Src in several human cancers, and describe how aberrantly expressed and activated Src is being investigated as a target for anticancer therapeutics.

4.3 SRC Family Kinases: Structure

All Src family members are proteins with an approximate molecular weight of 60 kD (range 56–62 kD), and all members have a uniform domain architecture (Fig. 4.2), consisting of a unique amino terminal end, involved in localization to the membrane and known as the SH_4 domain, a unique domain poorly conserved among SFKs, an SH_3 domain, involved in protein–protein interactions via binding to proline-rich P-X-X-P motifs, an SH_2 domain, involved in protein–protein interactions via binding to phosphorylated tyrosine residues, the tyrosine kinase domain, known as the SH_1 domain, which contains a tyrosine residue (Y419 in human Src) critical for Src activation, and a very short carboxy-terminal 'tail' region containing a regulatory sequence dominated by a tyrosine residue at position 530 in human Src, which, when phosphorylated, plays a major role in Src regulation (reviewed in Roskoski 2005, and see below). The crystal structure for Src has been determined, and illustrates, not unexpectedly, that all of the protein's domains play a role in the tertiary conformation, and thus activation state of SFKs (Xu et al. 1999, and see Regulation below).

As stated above, Src family members are all myristoylated (C14:0, tetradecanoic acid) on the penultimate glycine at the protein amino-terminus, a modification that enables the proteins to associate as peripheral membrane proteins with the inner leaflet of the cellular plasma membrane. The myristoyl fatty acid penetrates into the

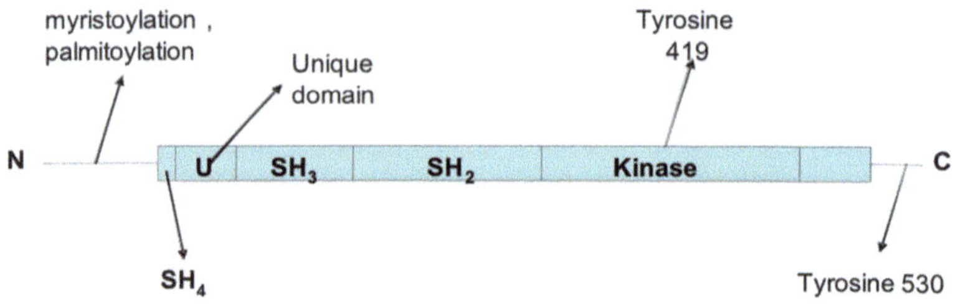

Fig. 4.2 Scematic structure of Src family kinases

acyl core of the membrane, aligning with the membrane phospholipid fatty acyl chains, providing an anchor for protein–membrane and protein–protein associations. In addition, seven members of the Src family, with the exceptions of Src itself and Blk, are palmitoylated (C16:0; hexadecanoic acid) on cysteine residues near the amino-terminal of the proteins. Palmitoylation is thought to be involved in protein–protein interactions, subcellular trafficking between cellular compartments, and, of course, in membrane association. However, since palmitoylation is reversible (while myristoylation is not), this posttranslational modification can be used by the cell in the dynamic regulation of subcellular location, and thus function (Basu 2004; Koegl et al. 1994).

4.4   SRC Family Kinases: Regulation

The specific activity of the tyrosine kinase of Src family members is regulated by intra- and intermolecular interactions, the former leading to a "closed" conformation with low kinase activity; the latter leading to an "open" conformation with increased specific activity of the kinase. The crystal structure of Src reveals that the intramolecular interactions required to maintain either conformation are complex and elegant, and are entirely dependent on the domain structure of the SFKs (Xu et al. 1999). In the closed configuration, the stabilization of the structure is maintained by phosphorylation of the Y530 residue, which then interacts with the Src SH_2 domain in an intramolecular association. Phosphorylation of the Y530 residue can be catalyzed by two extrinsic cellular PTKs, the C-terminal Src Kinase (*Csk*), or the Csk Homologue Kinase. In this inactive conformation, the A-loop helix of the protein, which contains the tyrosine at position 419 (in the human Src protein) that is phosphorylated upon activation, depends on the two-lobe structure of the tyrosine kinase SH_1 domain, and not only inhibits Y419 autophosphorylation, but also blocks Src kinase substrates from binding to the active pocket of the SH_1 domain (reviewed in Roskoski 2005).

Src assumes a more "open" conformation upon interaction with numerous proteins through either the Src SH-2 or SH-3 domain. This open conformation leads to dephosphorylation of the Y530 residue, which may be performed by a variety of cellular protein tyrosine phosphatases (PTPs), including PTP1B and SHP-1, although some studies have also implicated SHP-2, CD45, PTPα, PTPε, and PTPλ in this activity (reviewed in Roskoski 2005). Dephosphorylation of Y530, or displacement of the Src SH_2 and SH_3 domains from intramolecular interactions by competitive binding with numerous other proteins (e.g., Focal Adhesion Kinase, FAK, p130[CAS], or numerous growth factor receptors), disrupts the A-loop helix and exposes the Y419 to autophosphorylation, and/or phosphorylation at different sites by various cellular growth factor receptors (Roskoski 2005). As examples of the latter, ligand activation of a variety of cell surface receptors, including receptor PTKs (e.g., epidermal growth factor receptor, EGFR, PDGFR, c-Met, c-*erb*-B2/HER-2), and GPCR leads to molecular interactions with, and activation of, Src.

Once Y419 is phosphorylated, substrate binding to the kinase active site is optimized, the SH_2 and SH_3 domains are free to interact with extrinsic cellular protein-binding partners, and the Y530 regulatory residue is again available for phosphorylation. The process is thus dynamic, and partly dependent on the temporal phosphorylation or dephosphorylation of Src regulatory residues (reviewed in Roskoski 2005).

While the basic process of regulation of activity of Src and SFKs from a molecular point of view is established, the panoply of cellular proteins involved in this regulatory process is far from fully elucidated, and other activating mechanisms exist. In addition, there are other sites of phosphorylation and dephosphorylation on Src that affect protein–protein interactions. For example, although it is suspected that human PTP-BAS dephosphorylates the Y419 residue, inactivating Src, this process has not been experimentally shown to occur in human cells but is based on studies with a murine PTP analog (Roskoski 2005). In addition, although it has been shown that the platelet-derived growth factor (PDGF) receptor PTK can phosphorylate Src Y138 (Broome & Hunter 1997), this event has no known effect on Src activity or function, but likely affects the proteins with which Src associates. Further, PDGF and the c-*erb*-B2/HER-2 PTKs phosphorylate Src Y213 (Y215 in Humans), which activates Src, but the specific biologic relevance of these phosphorylations remains unknown (Stover et al. 1996; Vadlamudi et al. 2003). Finally, the Src kinase is a substrate for numerous Protein Serine/Threonine Kinases (PSTK), including Protein Kinase C, Protein Kinase A, and CDKs/cdc2, although it is only phosphorylation by the latter enzyme that appears to increase Src activity (Shenoy et al. 1992; Stover et al. 1994).

Since Src and other SFKs are integrally and vitally involved in numerous cellular functions, some are of great importance in human cancers, a complete understanding of the regulation of SFK function by intrinsic and extrinsic cellular protein mediators, and under which circumstances, is crucial to not only a complete understanding of those cellular functions, but how pharmaceutical intervention might allow alteration of those functions in the treatment of human diseases (Trevino et al. 2006).

4.5 Effects of Targeted Deletion of SFKs in Mice

The use of single gene mutation 'knockout' technology in mice has proved fruitful for the study of the physiological function of many genes. Functional deletions in mice of each of the SFKs have been generated. Some surprising findings have resulted from these studies, which are summarized as follows: functional deletions in Yes, Hck, Fgr, and Blk yield *no obvious abnormal mouse phenotype*; the Lyn functional deletion displayed impaired B-cell lymphocyte function and subsequent autoimmune dysfunction as well as developmental defects in the prostate; the Lck functional deletion displayed failure in T-cell lymphocyte development and impaired T-cell receptor function; the Fyn functional deletion displayed abnormal brain development and impaired memory; and the Src functional deletion *only* displayed osteopetrosis, a disease of bone where defective osteoclast activity results in an accumulation of bone (reviewed in Lowell & Soriano 1996). Given the ubiquitous expression pattern of Src, Fyn, and Yes in multiple cell lineages, and the role of Src in a variety of vital cellular pathways (Fig. 4.1), what is most surprising about these knockouts is their relatively limited effect on mouse development and function. On the other hand, the ubiquitous expression of several SFKs may lead to functional redundancy in which the expressed SFKs compensate for those functionally inactivated. Indeed, double mutant knockout mice with both Src and Yes mutated, or Src and Fyn mutated, have no distinguishable phenotypic feature, yet die at birth, supporting functional redundancy (Stein et al. 1994).

The remainder of this section focuses on the phenotype observed with Src knockouts. Functional deletion of the c-*src* gene leads to a bone disease known as osteopetrosis. These mice fail to display mature incisors, and thus cannot chew food and fail to thrive once weaned, unless kept on a soft-food diet (Soriano et al. 1991). Osteoclasts are bone-specific cells derived from the macrophage/monocyte lineage, which normally function to resorb bone material. They function in a balance with osteoblasts, which regenerate bone material. Osteoclast function requires both RANK (receptor activator of nuclear factor κB) and m-CSF (macrophage colony stimulating factor) function (Boyle et al. 2003; Yang et al. 1996). M-CSF acts as a ligand for its cellular receptor c-*fms*, which is a GFRTK found in the cytoplasmic membrane of osteoclasts and elsewhere (Ross 2006). Activation of c-*fms* further activates Src, which functions as described earlier to facilitate signaling through a variety of cellular pathways (Ross 2006). Src is expressed at high levels in osteoclasts (Thomas & Brugge 1997). Interestingly, in c-*src* −/− mice, osteoclasts are increased in number relative to "normal" mice (Boyce et al. 1992). Normal osteoclast function can be rescued in c-*src* −/− mice by transfer of normal osteoclasts from c-*src* +/+ donors (Lowe et al. 1993).

The molecular role(s) that Src plays in regulating osteoclast function has, to some extent, been determined by functionally inactivating additional genes. As seen earlier, if SFKs expressed in the same cells can compensate for loss of family members, then dual knockouts should yield valuable information as to how SFK family member compensate for the functions lost by other family members. In this light, it has been observed that the Hck and Fgr SFKs are also expressed at high levels in mouse osteoclasts. In fact, c-*hck* −/−; c-*src* −/− double-mutant knockouts display a significantly more severe osteopetrosis than the c-*src* −/− single mutant knockout. In c-*src* −/− single mutant knockouts, Hck protein expression, and presumably activity, are increased, which suggests that Hck compensates for the loss of Src. Finally, it has been observed that Src kinase activity is hyperactivated in osteoclasts (Lowell et al. 1996). These findings suggest that Src activity is crucial for osteoclast function, and that Hck only partially compensates for loss of Src, a point that has become of vital importance in an analysis of the role that Src plays in the pathogenesis of various human cancers, including prostate cancer, which frequently metastasizes to the bone. Ongoing clinical trials with SFK inhibitors reveal that bone turnover markers are generally reduced in responding patients, emphasizing that Src inhibition affects cells in the microenvironment that contribute to the vicious cycle of bone metastasis. Thus, inhibition of SFKs affects tumor cells, the microenvironment, and many of their interactions, a targeting strategy showing promise for therapy of prostate cancer that has metastasized to the bone (see below).

4.6 SRC and Human Cancers

Src expression and activity has been extensively studied in a wide variety of human cancers (reviewed in Summy & Gallick 2003). Src has been most extensively studied in human colorectal carcinoma (reviewed in Kopetz 2007), but an increasing literature strongly suggests that Src also plays an instrumental role in cancers of the prostate, breast, ovary, pancreas, bladder, lung, brain, and head and neck, as well as of melanoma and hematopoietic malignancies (reviewed in Summy & Gallick 2003).

Early findings correlated Src overexpression and abnormally high specific activity in colon cancers. Src protein expression and intrinsic tyrosine kinase activity are elevated in colon adenocarcinomas, or the cell lines generated from colon adenocarcinomas, relative to normal colonic epithelium (Bolen et al. 1987). Activation of Src occurs at an early stage, with polyps of high malignant potential and ulcerative colitis both showing increased Src-specific activity (Cartwright et al. 1994). Src activity further increases with disease stage, with liver metastases higher than primary tumors and secondary metastases to other sites even higher in activity. Elevated Src activity has been found to be a prognostic marker for all stages of colonic adenocarcinoma (Cartwright et al. 1994; Han et al. 1996; Allgayer et al. 2002). Only one report has suggested increased activity that may be due to rare mutations in advanced colon cancers (Irby et al. 1999); thus the large majority of the time Src activity almost certainly increases by nonmutagenic processes, perhaps protein–protein associations, as described above. This finding may help to explain why Src was, until recently, not vigorously pursued as an oncogene in human cancers, as the "oncogene addiction" theory is most applicable when activating mutations of these genes lead to constitutive deregulation of signaling pathways.

The role of Src overexpression and activation in human cancers, in the absence of genomic mutation, is now being extensively studied in a variety of human cancers. Indeed, Src may be the most important paradigm for the study of the increased activation of an enzymatically active molecule that is involved in so many vital cellular pathways playing a major role in oncogenic transformation and metastasis.

The relationship of aberrant SFK activation to human prostate cancer development, progression, and bone metastasis is receiving increasing attention. Two SFKs, Src and Lyn, have been found to be overexpressed in prostate cancer cell lines, and in a large percentage of prostate cancer tissues taken from patients (Goldenberg-Furmanov et al. 2004). Lyn has been associated with prostate tumor development, and its inhibition leads to decreased growth of prostate tumors in vitro and in vivo (Park et al. 2008). As in other solid tumors, Src activation affects migration and invasion, as well as tumor–stromal interactions (reviewed in Summy & Gallick 2003). The inhibition of Src also inhibits androgen-independent growth and metastasis (Lee et al. 2004). For these reasons, clinical trials of Src inhibitors are most advanced in metastatic prostate cancer (see below).

As stated earlier, Src is obviously not the only SFK activated in human cancers. The Yes SFK is also activated in a large fraction of colonic cancers (Han et al. 1996) and in melanoma (Loganzo et al. 1993), but is much less well studied. Interestingly, although the Src and Yes kinases can be coexpressed in a large fraction of *primary* colonic tumors, they are rarely if ever upregulated and activated in the same *metastatic* colonic tumors (Han et al. 1996), and Yes activation is a better marker of poor prognosis than Src in liver metastases (Han et al. 1996). Src and Yes do not perform the same roles in the pathogenesis of colon cancers (Park & Cartwright 1995). Indeed, upregulated Yes does not appear to be able to redundantly compensate for Src in c-*src* (−/−) gene knockouts (Ignelzi et al. 1994). The Lck SFK has also been found to be expressed in some colon carcinomas, higher in metastases, which is unusual since it is heretofore been solely an SFK of hematopoietic cells (Veillette et al. 1987).

Two additional mechanisms for hyperactivation of Src in human cancers demonstrate the complexity of Src regulation. First, aberrant overexpression and consequent increase in tyrosine phosphatase (PTP) activity, especially of those PTPs that have been shown to activate Src through dephosphorylation of Y530, e.g., PTP1B, PTPα, SHP-1, and SHP-2, can lead to constitutive Src activation (reviewed in Roskoski 2005). As an example, PTP1B overexpression in colon cancer cells reduced Y530 phosphorylation, increasing the tumorigenic potential of these cells in the absence of any significant change in Y419 phosphorylation (Zhu et al. 2007). A similar finding was observed in human breast cancer cells (Bjorge et al. 2000). Second, in many human cancers the aberrant constitutive overexpression and/or activation of GFPTKs or GPCR, and even steroid receptors can cause constitutive activation of Src (reviewed in Ishizawar & Parsons 2004). As an example, the (EGFR) tyrosine kinase, and the c-*erb*B-2 tyrosine kinase, both GFPTKs, can activate Src, and both are overexpressed and activated in breast cancers (reviewed in Johnson & Gallick 2007). In prostate cancer, a truncated version of the Kit GFPTK has been demonstrated in advanced prostate cancer tissues, and where found, also observed is increased expression and activation of Src (Paronetto et al. 2004). Finally, in colon cancers, the EGFR, c-*erb*B-2, and Met GFPTKs are all aberrantly expressed, and preferentially activate Src in highly metastatic cells (Mao et al. 1997). These two aspects of Src activation have been the subject of intense investigation in a variety of human cancers, and will undoubtedly lead to an improved understanding of the network of signaling proteins involved in tumor progression and strategies to inhibit them. As an example, Src phosphorylation of EGF-R on tyrosine 845 has been associated with resistance to EGF-R inhibitors; suggesting that cocktails of Src inhibitors *plus* EGF-R inhibitors may be of therapeutic benefit in some human tumors, a concept now being applied in clinical trials.

Most of the previously described studies concerning Src overexpression, hyperactivation, and cancer potential have been correlative in nature. However, the capability to up- or downregulate Src expression and activity provides a direct mechanism for a cause-and-effect analysis of the role of Src in human cancers. Early studies using nonspecific tyrosine kinase inhibitors, e.g., Herbimycin A, indicated that the inhibition of Src activity was decreased concurrent with loss of colon cancer cell growth in vitro (Garcia et al. 1991). More specific targeted approaches to shedding light on the role of Src have included the use of antisense Src technology, which, when used in human colon cancer cells to reduce the expression and activity of Src, caused very slow growing tumors, when compared to sense or vector controls.

This methodology also clearly illustrated that Yes, which was not downregulated by Src-specific antisense methods, does not play a redundant role in these cells, and was not able to rescue downregulated Src (Staley et al. 1997). Thus, constitutively activated SFKs in *tumor* cells may play distinct functions than those in *normal* cells, where functional redundancy has been observed.

More recently, additional roles for Src in tumor progression have been investigated, especially with respect to acquired chemoresistance and epithelial to mesenchymal transition (EMT). The last decade has seen a large number of reports in the scientific literature that indicate that aberrant Src activation leads to chemoresistance to therapeutic agents in cancers (reviewed in Shah & Gallick 2007). For example, the ectopic expression of constitutively activated v-*src* induced chemoresistance to cisplatin (Masumoto et al. 1999) and gefitinib (Qin et al. 2006), in gallbladder carcinomas. A different experimental approach showed the same phenomenon: pancreatic cancer cell lines which were induced to display increasing resistance to gemcitabine had correspondingly higher Src phosphorylation and activation in the absence of increased Src expression (Duxbury et al. 2004a). By using small interfering RNAs (siRNA) to Src, this same group showed that decreasing Src expression increased the sensitivity to gemcitabine (Duxbury et al. 2004b).

These findings are not limited to gallbladder or pancreatic carcinomas. Breast cancer cells that have become more resistant to tamoxifen display increased motility and invasion while simultaneously displaying increased Src activation (Morgan et al. 2009). In ovarian carcinomas, which display high Src expression and activity (Wiener et al. 2003), and which also usually develop chemoresistance rather rapidly, siRNA to Src regenerated sensitivity to paclitaxel and cisplatin (Chen et al. 2005). Similar results were observed in colonic adenocarcinoma cells (Griffiths et al. 2004).

The phenomenon of EMT of tumor cells may explain how cancer cells invade and metastasize. As epithelial cells acquire properties of fibroblasts, they lose cell-to-cell contact inhibition, a hallmark of cancer cells, and at the same time lose E-Cadherin expression (Guarino et al. 2007; Hay 1995). E-Cadherin functions in cell-to-cell contacts via adherens junctions, also involving integrins, so the loss of E-Cadherin facilitates cell migration, an obvious first step in invasion and metastasis (reviewed in Guarino et al. 2007). In colon cancer cells, E-Cadherin is decreased in expression when Src is overexpressed (Avizienyte et al. 2002). When EMT occurs, as E-Cadherin-mediated cell-to-cell contacts are reduced, integrin-associated adherens junctions are increased, an event that requires intact Src catalytic activity, and intact SH_2 and SH_3 domains of Src (Guarino et al. 2007; Avizienyte et al. 2002; Coluccia et al. 2006).

Is Src overexpression and/or hyperactivation a cause of chemoresistance and EMT in cancer cells? An ever-growing body of evidence indicates that Src activation coordinates with increased metastatic potential and increased chemoresistance, but the molecular mechanisms by which this occurs are only now being clarified. Clearly, if Src activation is a causative factor in EMT and chemoresistance, pharmaceutical intervention aimed at reducing Src activity may be important in sensitizing tumor cells to standard chemotherapeutic agents to which they have become resistant.

4.7 SRC as a Target for Pharmaceutical Intervention Against Cancer

Currently, several SFK inhibitors are in clinical trial for solid tumors. These include bosutinib, saracatinib, dasatinib, all of which are competitive inhibitors of ATP binding, and KX2-391, which affects Src substrate binding. As yet, their success cannot be predicted though a consensus is emerging that such inhibitors are of little value as single agents (reviewed in Kopetz et al. 2007; Summy & Gallick 2006).

Given the conservation of tyrosine kinase active sites, most SFK competitive inhibitors are also Abl PTK inhibitors. Thus, dasatinib is FDA approved for Gleevec-resistant chronic myelogenous leukemia; its success due to its ability to inhibit the BCR-ABL fusion protein resulting from a 9:22 chromosomal translocation. Dasatinib and other SFK inhibitors have shown more limited effects as antitumor agents in most other trials. These results were predictable from single-agent pre-clinical studies, which have shown in general that the inhibitors have a very limited effect on cellular proliferation or tumor growth in animals, but are very potent anti-metastatic agents (Kopetz et al. 2007; Summy & Gallick 2006).

Combination therapies with SFK inhibitors have shown more promise. Many of these combinations use both targeted agents (such as EGF-R or c-*erb*B-2/HER-2 monoclonal antibody inhibitors and/or antiangiogenic agents) with traditional chemotherapy. While many of the combination therapies are empirical, adding Src inhibitors to other agents approved or in clinical trial for solid tumors, current successes hint at the settings in which SFK inhibitors might become part of the standard arsenal of anticancer therapy. As discussed above, the activation of Src that can induce chemoresistance, and the ability of Src inhibitors to resensitize tumor cells to these agents suggest one possible scenario where the use of SFK inhibitors might become part of standard of care for some tumors. Thus, it seems at this time the best promise for Src inhibitors will be as sensitizers to chemotherapy, radiation therapy, and some targeted therapies.

As the role of the tumor microenvironment in promoting tumor growth is becoming better understood, perhaps the most promising use of SFK inhibitors will be in settings in which they affect both tumor progression and the microenvironment. Prostate cancer metastasis to the bone is the current paradigm for targeting both tumor and microenvironment. The "vicious cycle" of bone metastasis in which tumor cells stimulate bone resorption/formation, causing the release of factors that contribute to tumor growth, stimulating more bone turnover (Fizazi 2007), requires that successful therapies target both tumor and microenvironment (Efstathiou & Logothetis 2010). As discussed above, Src inhibition not only affects prostate tumor growth and invasion, but also osteoclast function, thus potentially interrupting this vicious cycle. Here again, SFK inhibitors such as dasatinib show little promise as a single agent. However, in combination with docetaxel, declining PSA is associated with a decline in bone turnover markers in responding patients, a result so encouraging that Phase 2 trials on this combination are completed (Araujo et al. 2011), with a subset of patients showing long-term survival without increase in prostate-specific antigen. These very promising results led to a recently completed Phase 3 trial (Araujo et al. 2009).

In summary, the study of nonreceptor kinases of the Src family began with esoteric models of initially unclear importance that eventually demonstrated the importance of aberrant activation of proto-oncogene protein products in human cancer. More recently, development of SFK inhibitors coupled with translational 'bench-to-bedside' approaches has altered thinking as to how tumor therapy must proceed, i.e., targeting both tumor and microenvironment. Whether SFK inhibitors become *standard of care* remains uncertain, but continued study of this most venerable family of proteins most assuredly will lead to new principles in cancer biology and cancer therapy. The future of research on this small nine-member kinase family is likely to be as promising as the remarkable discoveries made over the last century.

References

Allgayer H, Boyd DD, Heiss MM, Abdalla EK, Curley SA, Gallick GE (2002) Activation of Src kinase in primary colorectal carcinoma: an indicator of poor clinical prognosis. Cancer 94:344–351

Araujo J, Armstrong AJ, Braud EL, Posadas E, Lonberg M, Gallick GE Trudel GC, Paliwal P, Agrawal S, and Logothetis CJ (2009) Dasatinib and docetaxel combination treatment for patients with castration-resistance progressive prostate cancer: a phase I/II study (CA180086). J Clin Oncol 27(15S):249s [abstract 5061]

Araujo JC, Mathew P, Armstrong AJ, Braud EL, Posadas E, Lonberg M, Gallick GE, Trudel GC, Paliwal P, Agrawal S, Logothetis SJ (2011) Dasatinib combined with docetaxel for castration-resistant prostate cancer: results from a phase 1/2 study. Cancer. (in press)

Avizienyte E, Wyke AW, Jones RJ, McLean GW, Westhoff MA, Brunton VG, Frame MC (2002) Src-induced deregulation of E-cadherin in colon cancer cells requires integrin signalling. Nat Cell Biol 4:632–638

Basu J (2004) Protein palmitoylation and dynamic modulation of protein function. Current Science 87(2):212–217

Bjorge JD, Pang A, Fujita DJ (2000) Identification of protein-tyrosine phosphatase 1B as the major tyrosine phosphatase activity capable of dephosphorylating and activating c-Src in several human breast cancer cell lines. J Biol Chem 275(52):41439–41446

Bolen JB, Veillette A, Schwartz AM, DeSeau V, Rosen N (1987) Activation of pp 60c-src protein kinase activity in human colon carcinoma. Proc Natl Acad Sci U S A 84:2251–2255

Boyce BF, Yoneda T, Lowe C, Soriano P, Mundy GR (1992) Requirement of pp 60c-src expression for osteoclasts to form ruffled borders and resorb bone in mice. J Clin Invest 90:1622–1627

Boyle WJ, Simonet WS, Lacey DL (2003) Osteoclast differentiation and activation. Nature 423:337–342

Broome MA, Hunter T (1997) The PDGF receptor phosphorylates Tyr138 in the c-Src SH3 domain in vivo reducing peptide ligand binding. Oncogene 14:17–34

Cartwright CA, Coad CA, Egbert BM (1994) Elevated c-Src tyrosine kinase activity in premalignant epithelia of ulcerative colitis. J Clin Invest 93:509–515

Chen T, Pengetnze Y, Taylor CC (2005) Src inhibition enhances paclitaxel cytotoxicity in ovarian cancer cells by caspase-9-independent activation of caspase-3. Mol Cancer Ther 4:217–224

Coluccia AM, Benati D, Dekhil H, DeFilippo A, Lan C, Gambacorti-Passerini C (2006) SKI-606 decreases growth and motility of colorectal cancer cells by preventing pp 60 (c-Src)-dependent tyrosine phosphorylation of beta-catenin and its nuclear signaling. Cancer Res 66:2279–2286

Demory ML, Boerner JL, Davidson R, Faust W, Miyake T, Lee I, Hüttemann M, Douglas R, Haddad G, Parsons SJ (2009) Epidermal growth factor receptor translocation to the mitochondria: regulation and effect. J Biol Chem 284:36592–36604

Duxbury MS, Ito H, Zinner MJ, Ashley SW, Whang EE (2004a) Inhibition of SRC tyrosine kinase impairs inherent and acquired gemcitabine resistance in human pancreatic adenocarcinoma cells. Clin Cancer Res 10:2307–2318

Duxbury MS, Ito H, Zinner MJ, Ashley SW, Whang EE (2004b) siRNA directed against c-Src enhances pancreatic adenocarcinoma cell gemcitabine chemosensitivity. J Am Coll Surg 198:953–959

Efstathiou E, Logothetis CJ (2010) A new therapy paradigm for prostate cancer founded on clinical observations. Clin Cancer Res 16:1100–1107

Fizazi K (2007) The role of Src in prostate cancer. Ann Oncol 18:1765–1773

Garcia R, Parikh NU, Saya H, Gallick GE (1991) Effect of herbimycin a on growth and pp 60c-src activity in human colon tumor cell lines. Oncogene 6:1983–1989

Goldenberg-Furmanov M, Stein I, Pikarsky E, Rubin H, Kasem S, Wygoda M, Weinstein I, Reuveni H, Ben-Sasson SA (2004) Lyn is a target gene for prostate cancer; sequence-based inhibition induces regression of human tumor xenografts. Cancer Res 64:1058–1066

Griffiths GJ, Koh MY, Brunton VG, Cawthorne C, Reeves NA, Greaves M, Tilby MJ, Pearson DG, Ottley CJ, Workman P, Frame MC, Dive C (2004) Expression of kinase-defective mutants of c-Src in human metastatic colon cancer cells decreases Bcl-xl and increases oxaliplatin- and Fas-induced apoptosis. J Biol Chem 279:46113–46121

Guarino M, Rubino B, Ballabio G (2007) The role of epithelial-mesenchymal transition in cancer pathology. Pathology 39:305–318

Han NM, Curley SA, Gallick GE (1996) Differential activation of pp 60(c-src) and pp62 (c-yes) in human colorectal carcinoma liver metastases. Clin Cancer Res 8:1397–1404

Hay ED (1995) An overview of epithelio-mesenchymal transformation. Acta Anat 154:8–20

Hunter T, Sefton BM (1980) Transforming gene product of Rous sarcoma virus phosphorylates tyrosine. Proc Natl Acad Sci U S A 77:1311–1315

Ignelzi MA Jr, Miller DR, Soriano P, Maness PF (1994) Impaired neurite outgrowth of Src-minus cerebellar neurons on the cell adhesion molecular L1. Neuron 12:873–884

Irby RB, Mao W, Coppola D, Kang J, Loubeau JM, Trudeau W, Karl R, Fujita DJ, Jove R, Yeatman TJ (1999) Activating SRC mutation in a subset of advanced human colon cancers. Nat Genet 21:187–190

Ishizawar R, Parsons SJ (2004) C Src and cooperating partners in human cancer. Cancer Cell 6:209–214

Johnson FM, Gallick GE (2007) Src family nonreceptor tyrosine kinases as molecular targets for cancer therapy. Anticancer Agents Med Chem 7:651–659

Koegl M, Zlatkine P, Ley SC, Courtneidge SA, and Magee AI. (1994) Palmitoylation of multiple Src-family kinases at a homologous N-terminal motif. Biochem J. 303 (pt3): 749-53. [Mol Cell. 1999; 3(5):629–38]

Kopetz S (2007) Targeting Src and epidermal growth factor receptor in colorectal cancer: rationale and progress into the clinic. Gastrointest Canc Res 1(4 Suppl 2):S37–S41

Kopetz S, Shah AN, Gallick GE (2007) Src continues aging: current and future clinical directions. Clin Cancer Res 13(24):7232–7236

Lee LF, Louie MC, Desai SJ, Yang J, Chen HW, Evans CP, Kung HJ (2004) Interleukin-8 confers androgen-independent growth and migration of LNCaP: differential effects of tyrosine kinases Src and FAK. Oncogene 23:2197–2205

Loganzo F Jr, Dosik JS, Zhao Y, Vidal MJ, Nanus DM, Sudol M, Albino AP (1993) Elevated expression of protein tyrosine kinase c-Yes, but not c-Src, in human malignant melanoma. Oncogene 8:2367–2644

Lowe C, Yoneda T, Boyce BF, Chen H, Mundy GR, Soriano P (1993) Osteopetrosis in Src-deficient mice is due to an autonomous defect of osteoclasts. Proc Natl Acad Sci U S A 90:4485–4489

Lowell CA, Soriano P (1996) Knockouts of Src-family kinases: stiff bones, wimpy T cells, and bad memories. Genes Dev 10:1845–1857

Lowell CA, Niwa M, Soriano P, Varmus HE (1996) Deficiency of the Hck and Src tyrosine kinases results in extreme levels of extramedullary hematopoiesis. Blood 87:1780–1792

Mao W, Irby R, Coppola D, Fu L, Wloch M, Turneer J, Yu H, Garcia R, Jove R, Yeatman TJ (1997) Activation of c-Src by receptor tyrosine kinases in human colon cancer cells with high metastatic potential. Oncogene 15:3083–3090

Martin GS (1970) Rous sarcoma virus: a function required for the maintenance of the transformed state. Nature 227:1021–1023

Masumoto N, Nakaon S, Fujishima H, Kohno K, Niho Y (1999) v-src induces cisplatin resistance by increasing the repair of cisplatin-DNA interstrand cross-links in human gallbladder adenocarcinoma cells. Int J Cancer 80:731–737

Morgan L, Gee J, Pumford S, Farrow L, Finlay P, Robertson J, Ellis I, Kawakatsu H, Nicholson R, Hiscox S (2009) Elevated Src kinase activity attenuates tamoxifen response in vitro and is associated with a poor prognosis clinically. Cancer Biol Ther 8:1550–1558

Park J, Cartwright CA (1995) Src activity increases and Yes activity decreases during mitosis of human colon carcinoma cells. Mol Cell Biol 15:2374–2382

Park SI, Zhang J, Phillips KA, Araujo JC, Najjar AM, Volgin AY, Gelovani JG, Kim SJ, Wang Z, Gallick GE (2008) Targeting SRC family kinases inhibits growth and lymph node metastases of prostate cancer in an orthotopic nude mouse model. Cancer Res 68:3323–3333

Paronetto MP, Farini D, Sammarco I, Maturo G, vespasiani G, Geremia R, Rossi P, Sette C (2004) Expression of a truncated form of the c-Kit tyrosine kinase receptor and activation of Src kinase in human prostate cancer. Am J Pathol 164:1243–1251

Qin B, Ariyama H, Baba E, Tanaka R, Kusaba H, Harada M, Nakano S (2006) Activated Src and Ras induced gefitinib resistance by activation of signaling pathways downstream of epidermal growth factor receptor in human gallbladder adenocarcinoma cells. Cancer Chemother Pharmacol 58:577–584

Roskoski R Jr (2005) Src kinase regulation by phosphorylation and dephosphorylation. Biochem Biophys Res Commun 331:1–14

Ross FP (2006) M-CSF, c-Fms, and signaling in osteoclasts and their precursors. Ann N Y Acad Sci 1068:110–116

Rous P (1911) A sarcoma of the fowl transmissible by an agent separable from the tumor cells. J Exp Med 13:397–411

Shah AN, Gallick GE (2007) Src, chemoresistance and epithelial to mesenchymal transition: are they related? Anticancer Drugs 18(4):371–375

Shenoy S, Chackalaparampil I, Bagrodia S, Lin PH, Shalloway D (1992) Role of p34cdc2-mediated phosphorylations in two-step activation of pp 60c-src during mitosis. Proc Natl Acad Sci U S A 89:7237–7241

Soriano P, Montgomery C, Geske R, Bradley A (1991) Targeted disruption of the c-Src protooncogene leads to osteopetrosis in mice. Cell 64(4):693–702

Staley CA, Parikh NU, Gallick GE (1997) Decreased tumorigenicity of a human colon adenocarcinoma cell line by an antisense expression vector specific for c-Src. Cell Growth Differ 8:269–274

Stehelin D, Varmus HE, Bishop JM, Vogt PK (1976) DNA related to the transforming gene(s) of avian sarcoma viruses is present in normal avian DNA. Nature 260:170–173

Stein PL, Vogel H, Soriano P (1994) Combined deficiencies of Src, Fyn, and Yes tyrosine kinases in mutant mice. Genes Dev 8:1999–2007

Stover DR, Liebetanz J, Lydon NB (1994) Cdc2-mediated modulation of the pp 60c-src activity. J Biol Chem 269:26885–26889

Stover DR, Furet P, Lydon NB (1996) Modulation of the SH2 binding specificity and kinase activity of Src by tyrosine phosphorylation within its SH2 domain. J Biol Chem 271:12481–12487

Summy JM, Gallick GE (2003) Src family kinases in tumor progression and metastasis. Cancer Metastasis Rev 22:337–358

Summy JM, Gallick GE (2006) Treatment for advanced tumors: Src reclaims center stage. Clin Cancer Res 12(5):1398–1401

Thomas SM, Brugge JS (1997) Cellular functions regulated by Src family kinases. Annu Rev Cell Dev Biol 13:513–609

Trevino JG, Summy JM, Gallick GE (2006) Src inhibitors as potential therapeutic agents for human cancers. Mini Rev Med Chem 6:681–687

Vadlamudi RK, Sahin AA, Adam L, Wang RA, Kumar R (2003) Heregulin and HER2 signaling selectively Activates c-Src phosphorylation at tyrosine 215. FEBS Lett 543:76–80

Veillette A, Foss FM, Sausville EA, Bolen JB, Rosen N (1987) Expression of the lck tyrosine kinase gene in human colon carcinoma and other non-lymphoid human tumor cell lines. Oncogene Res 1:357–374

Wiener JR, Windham TC, Estrella VC, Parikh NU, Thall PF, Deavers MT, Bast RC, Mills GB, Gallick GE (2003) Activated SRC protein tyrosine kinase is overexpressed in late-stage human ovarian cancers. Gynecol Oncol 88:73–79

Xu W, Doshi A, Lei M, Eck MJ, Harrison SC (1999) Crystal structures of c-Src reveal features of its autoinhibitory mechanism. Mol Cell 3(5):629–638

Yang S, Zhang Y, Rodriguez RM, Ries WL, Key LL Jr (1996) Functions of the M-CSF receptor on osteoclasts. Bone 18:355–360

Zhu S, Bjorge JD, Fujita DJ (2007) PTP1B contributes to the oncogenic properties of colon cancer cells through Src activation. Cancer Res 67(21):10129–10137

Chapter 5
The Hedgehog Signaling Pathway in Cancer Pathogenesis and Therapy

Margaret A. Read and Vito J. Palombella

5.1   Introduction

The discovery of the Hedgehog (Hh) pathway by Nüsslein-Volhard and Wieschaus (1980) was recognized by a Nobel Prize in 1995. Their groundbreaking mutational analysis of genes in *Drosophila* that control segmentation and polarity elucidated a pathway that, when mutated, resulted in larvae with spiculated cuticles on their skin, resembling the spines of a hedgehog. Subsequent identification of the specific gene products revealed a unique signaling pathway with related orthologs in vertebrate organisms (reviewed in Jiang and Hui 2008). The ability to decipher Hh signaling pathways has benefited from genetics in model systems, with a heavy focus in *Drosophila* where the pathway was first discovered. Signaling through the Hh pathway involves two transmembrane proteins, Patched (Ptc) and Smoothened (Smo), and is regulated by the absence or presence of Hh ligands (Fig. 5.1). In most adult cells, Hh ligand is not present, and Ptc functions to repress the activity of Smo, keeping the pathway inactive. Upon binding of Hh ligand to Ptc, Smo inhibition is relieved. Derepression of Smo triggers a signal transduction cascade that activates Gli transcription factors resulting in expression of target genes that regulate cellular differentiation, migration, proliferation, and survival (Jiang and Hui 2008). Following an overview of signaling through the Hh pathway, this chapter will focus on how the pathway impacts multiple aspects of tumor growth and survival, through both Hh ligand-independent and ligand-dependent mechanisms, and provide a biological rationale for cancer treatment strategies using inhibitors of the Hh pathway.

M.A. Read (✉) • V.J. Palombella
Infinity Pharmaceuticals, Inc., Cambridge, MA 02139, USA
e-mail: Margaret.read@infi.com

D.A. Frank (ed.), *Signaling Pathways in Cancer Pathogenesis and Therapy*,
DOI 10.1007/978-1-4614-1216-8_5, © Springer Science+Business Media, LLC 2012

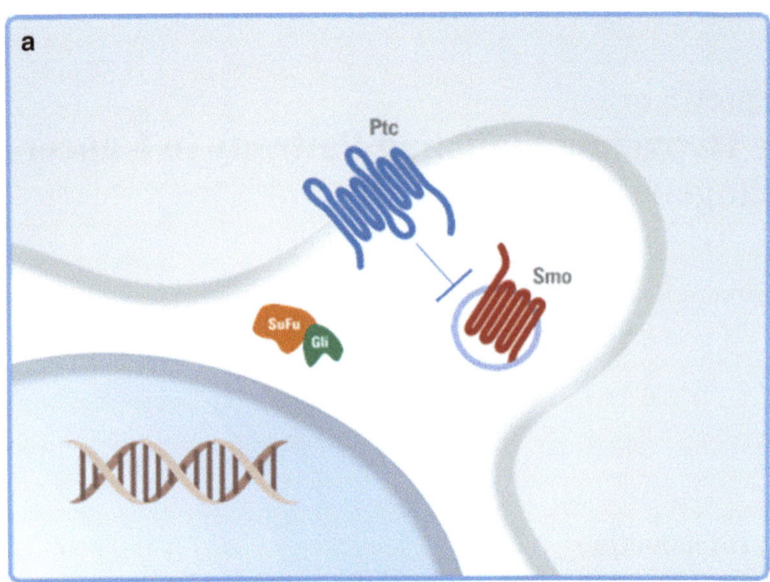

Fig. 5.1 The Hedgehog Pathway Malignant activation of the Hh pathway plays a key role in tumorigenesis and growth of certain cancers. This pathway involves two cell-membrane proteins, Patched (Ptc) and Smoothened (Smo), and is regulated by the absence or presence of Hh ligand. (**a**) The Hh signal transduction pathway plays a critical role in cell differentiation and patterning during development, but is inactive in most adult cells. In the absence of Hh ligand, the 12-transmembrane receptor Ptc is localized in the primary cilia and inhibits the activity of the GPCR-like Smo which is sequestered within vesicles in the cytosol. Inhibition of Smo ensures that Gli transcription factors, which activate target genes, are held in an inactive form via a complex that contains Suppressor of Fused (SuFu). (**b**) In certain cancers, such as basal cell carcinomas and some medulloblastomas, malignant activation of the Hh pathway is ligand-independent, with Hh signaling activated by genetic mutation. Mutations in Ptc can enable localization of Smo to the cilia and activation of the Hh signal transduction cascade. Gli transcription factors are released from the complex with SuFu and translocate to the nucleus. In the nucleus, the Gli transcription factors activate expression of genes that promote tumor growth and survival, including regulators of the cell cycle, differentiation, and apoptosis, such as Gli1, Cyclin D1 and Cyclin D2, Myc, and Bcl-2. (**c**) In ligand-dependent cancers, such as chondrosarcoma and osteosarcoma, Hh ligand binds to Ptc on the tumor cell. In the presence of Hh ligand, Ptc moves out of the primary cilia, relieving Ptc-mediated inhibition of Smo. This mode of ligand-dependent signaling may also occur in progenitor cells in certain cancers, including chronic myelogenous leukemia (CML), chronic lymphocytic leukemia (CLL), acute lymphocytic leukemia (ALL), non-Hodgkin's lymphoma (NHL), multiple myeloma (MM), and small cell lung cancer (SCLC), and may be responsible for minimal residual disease following treatment with chemotherapeutic and targeted agents. (**d**) In other cancers, such as pancreatic cancer, malignant activation of the Hh pathway is also ligand-dependent. Hh ligand produced by the tumor cells acts through Ptc on stromal cells in the tumor microenvironment. This mode of tumor-stroma paracrine signaling provides support for tumor growth and survival through mechanisms originating in the stromal cells. In pancreatic cancer models, inhibition of Hh signaling within the tumor microenvironment depletes the desmoplastic stroma, increases the vascularity of the tumor, and renders the tumor more accessible to chemotherapy. Paracrine signaling may also be responsible for malignant activation of the Hh pathway in other tumor types. (**e**) Hh pathway inhibitors that directly block the activity of Smo are currently being investigated in clinical trials. Because Smo plays a critical role in malignant activation of the Hh pathway, Smo may be a target for the management of a broad range of cancers. IPI-926 is a potent, orally delivered small molecule that is currently being investigated as an inhibitor of Smo in both ligand-dependent and independent settings

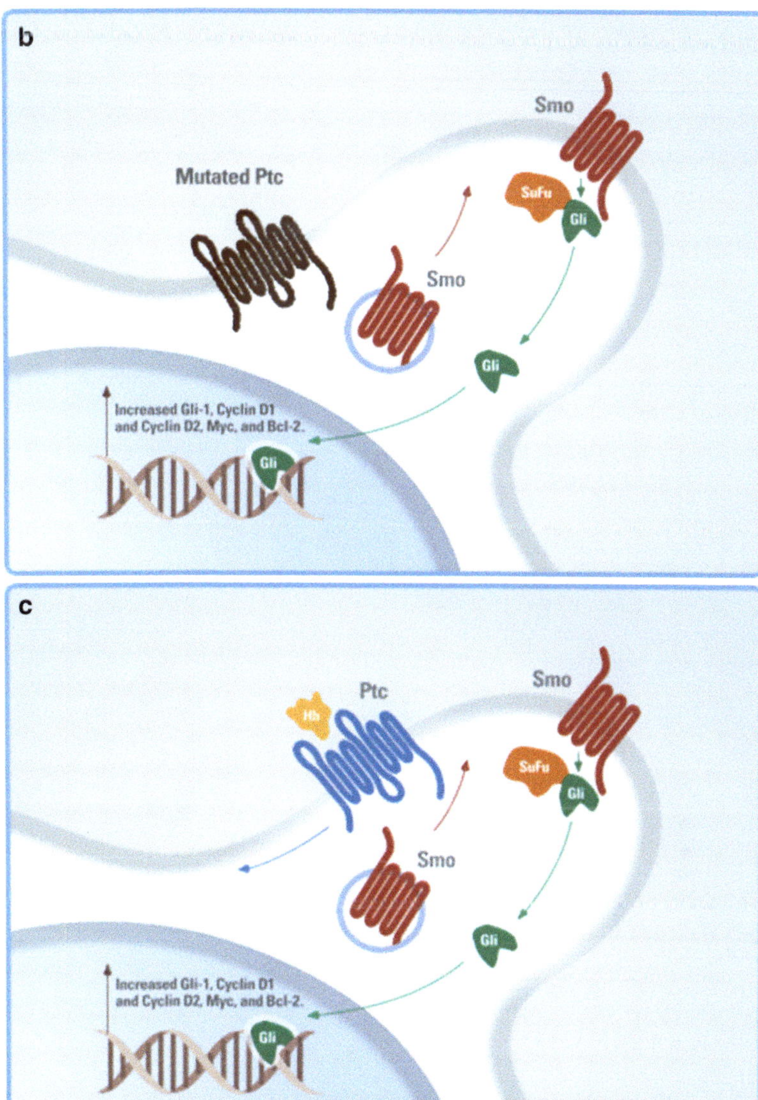

Fig. 5.1 (continued)

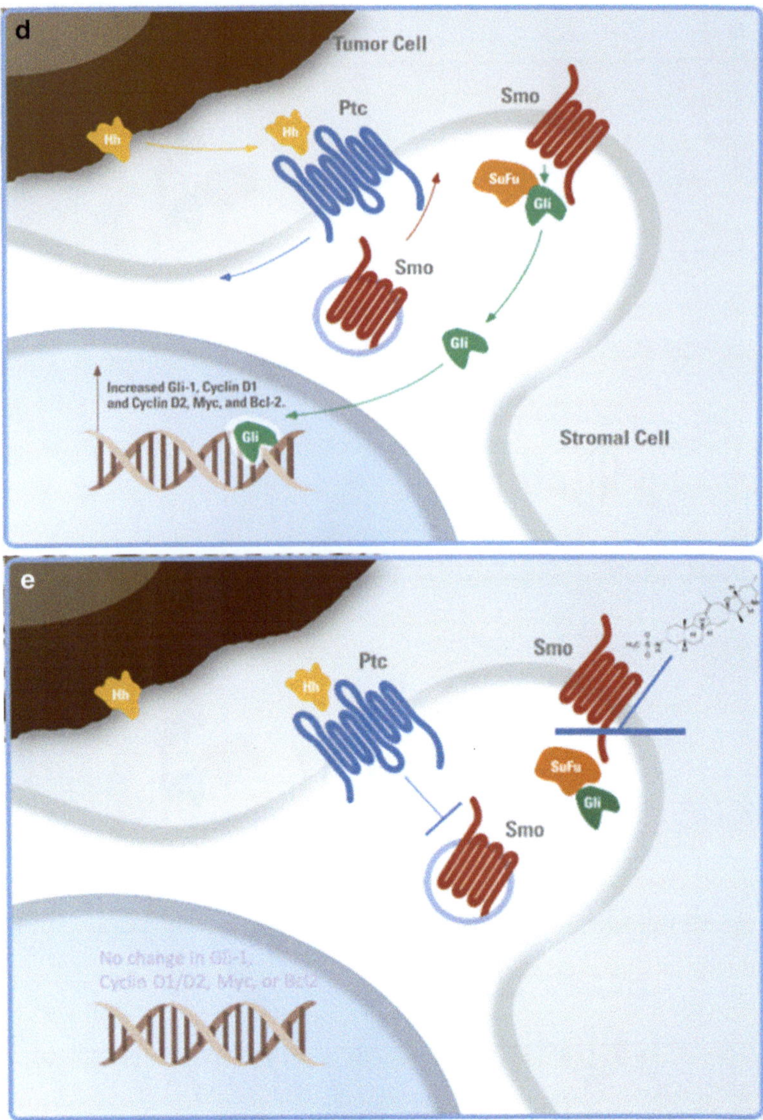

Fig. 5.1 (continued)

5.2   Signaling Through the Hedgehog Pathway

Important links between *Drosophila* genetics and vertebrate biology lead to breakthroughs in our understanding of the Hh pathway. One such link was the discovery of the plant-derived alkaloid cyclopamine that is produced by the corn lily plant (*Veratrum californicum*). This steroidal alkaloid was discovered through its teratogenic activity in developing lamb fetuses due to maternal ingestion of corn lily plants (Keeler 1968). Of note, the maternal ewes do not suffer ill effects from ingestion of the plants or cyclopamine, with birth defects being confined to a specific window of time during fetal development (Welch et al. 2009). Cyclopamine was subsequently found to antagonize the Hh pathway (Cooper et al. 1998; Incardona et al. 1998) and to exert its inhibitory effects by binding to Smo (Taipale et al. 2000; Chen et al. 2002). The natural product cyclopamine, while not active against *Drosophila* Smo, has served as a powerful tool to help understand the role of the Hh pathway in many aspects of mammalian physiology and disease. Following the discovery of Hh in *Drosophila*, mutations generated in the Hh pathway in vertebrates were shown to result in animals with cyclopic features (Chiang et al. 1996; Belloni et al. 1996). These findings were substantiated in humans, where mutations in Sonic (Shh) were linked to holoprosencepaly, which can include cyclopic features (Roessler et al. 1996). Thus, inhibitors of the Hh pathway, whether derived from cyclopamine or not, would be expected to impact embryogenesis.

Since these primary discoveries, the Hh pathway has received increased attention not only for its role in regulating embryonic organogenesis, but also as an oncogenic pathway that is involved in many human cancers. Malignant activation of the Hh pathway through Smo can promote growth of certain cancers via three modes: (1) ligand-independent signaling that is activated by genetic mutation; (2) ligand-dependent signaling to tumor cells; and (3) ligand-dependent signaling between the tumor and the microenvironment (Fig. 5.1). Malignant activation of the Hh pathway appears to be involved in both the initiation of cancer and tumor growth, survival, and metastases.

5.2.1   Activation of Smo: GPCR Activity and Requirement for Primary Cilia

While there is a high degree of conservation in Hh pathway components between invertebrates and higher organisms, not all aspects of signaling are conserved between the two systems. For example, in *Drosophila*, there is one Hh ligand, but in vertebrates, there is a family of Hh ligands: Indian (Ihh), Sonic (Shh), and Desert hedgehog (Dhh). Hh signaling is rarely maintained in cultured mammalian cells that have been propagated in the presence of high serum concentrations (Sasai et al. 2006). This finding coupled with a paucity of antibody reagents to specific mammalian pathway components has resulted in Hh pathway signaling being considerably less well-understood in mammalian cells. After a Hh ligand binds Ptc, the precise biochemical events that are involved in Hh-mediated activation of mammalian Smo are still being elucidated. However, two aspects of Smo activation are gaining acceptance and are likely to be interdependent. First, Smo bears homology to another seven-transmembrane protein, the guanine nucleotide-binding protein-coupled receptor (GPCR) Frizzled (Alcedo et al. 1996). Evidence that mammalian Smo functions as a GPCR includes a couple of key characteristics of GPCR-mediated signal transduction. Upon activation, Smo associates with β-arrestin 2 (βarr2) in a heterotrimeric GPCR kinase 2 (Grk2)-dependent manner (Chen et al. 2004). In some mammalian cells, signaling through Smo can also be coupled to G_i and modulation of cAMP (Ogden et al. 2008). A second aspect of Smo signaling involves sensing through a specialized form of cilia. Primary nonmotile cilia are essential for transduction of the Hh signal in mammalian cells. Rohatgi et al. (2007) showed how Hh pathway components dynamically traffic through the primary cilium using intraflagellar transport proteins (IFT) and that this trafficking is required for signal transduction. In the absence of Hh ligand, Ptc localizes to cilia and inhibits Smo by preventing its accumulation within cilia. When Shh binds to Ptc, Ptc leaves the cilia and is internalized through endosomes, leading to accumulation of Smo in primary cilia and activation of signaling. Accumulation of Smo in cilia has been linked to its role as a GPCR in that βarr2 is required for the association of Smo with a member of the IFT transport machinery, Kif3, which is a subunit of the kinesin-2 motor complex. This association of Smo with Kif3 is required for Smo activation of Gli (Kovacs et al. 2008). While transport of Smo to cilia is necessary for activation, recent studies indicate that a second undefined step is necessary to fully activate downstream events leading to target gene induction (Rohatgi et al. 2009). Thus, Shh ligand is first sensed in the primary cilia, and from there, signals are transduced via Smo that lead to activation of the Hh pathway.

5.2.2 Gli Regulation and Downstream Targets

Whereas in *Drosophila* there is one Gli-like protein, called Cubitus interruptus (CI), in mammalian cells, there is a family of Gli transcription factors, Gli1, Gli2, and Gli3 (reviewed in Jiang and Hui 2008). In the absence of Hh ligand, Gli3 is processed to a repressor form and keeps expression of Hh pathway genes in check. Upon derepression of Smo, Gli1 and 2 translocate to the nucleus where they primarily function as transcriptional activators, and Gli3 is no longer processed to a repressor form. Gli proteins can activate transcription of many genes involved in growth and development and also act as positive feedback regulators of the Hh pathway by inducing Gli1 transcription, making Gli1 gene expression one of the most reliable and robust measures of Hh pathway activation. A number of negative modulators such as Suppressor of fused (SuFu), Hh interacting protein-1 (Hhip1), and Ptc participate in negative feedback loops, ensuring that Gli function is tightly regulated. Gli1 and 2 regulate expression of a number of genes important for cell differentiation, migration, proliferation, and survival, including Bcl-2 (Regl et al. 2004), n-Myc (Oliver et al. 2003; Mill et al. 2005), cyclin-D1, cyclin-D2, plakoglobin, IGFBP6 (Yoon et al. 2002), IGFBP3 (Yu et al. 2009a), sFRP-1 (He et al. 2006), follistatin (Eichberger et al. 2008), snail (Li et al. 2006), and osteopontin (Das et al. 2009).

5.2.3 Role of Cilia in Oncogenesis

In studies in murine models of cancer that are driven by Hh ligand-independent activating mutations, Wong et al. (2009) and Han et al. (2009) found that the requirement for primary cilia depends upon where in the Hh pathway the oncogenic mutations occur. When mutations occur at the receptor level (Ptc or Smo), primary cilia are required for Hh signaling and subsequent tumor development. In contrast, when activating mutations are introduced downstream of Ptc and Smo (e.g., exogenous expression of activated Gli2), Hh signaling and tumor development can occur in the absence of functional primary cilia. In fact, a surprising observation was made in the context of exogenous expression of an activated form of Gli2; the presence of primary cilia actually repressed tumor formation, possibly because cilia are required for the generation of the Gli3 repressor (Liu et al. 2005). These findings further emphasize that the loss of the Gli3 repressor is important for full Hh pathway activation. For tumors that are driven by activating mutations in the Hh pathway, such as basal cell carcinoma and medulloblastoma (see below), primary cilia may serve as a biomarker for selection of tumors that will respond to inhibitors that exert their effects at the level of Smo and Ptc.

5.3 Malignant Activation of the Hh Pathway

5.3.1 Ligand-Independent Activation

The first link between the Hh pathway and cancer came from the discovery that Gorlin's syndrome, an inherited condition, is due to an autosomal loss of the gene for Ptc (reviewed in Bale and Yu 2001). Children with this condition have multiple physical defects and a predisposition for cancers including medulloblastoma and basal cell carcinoma. In addition, analysis of tumor tissue from sporadic BCC and medulloblastoma patients shows a high incidence of hyperactivation of the Hh pathway, demonstrated by high levels of Gli1 expression and inactivating mutations in Ptc as well as activating mutations in Smo (Hahn et al. 1996). Studies in transgenic murine models have confirmed that uncontrolled activation of the Hh pathway is an early event in the formation of these tumor types (Xie et al. 1997; Goodrich et al. 1997). Thus, Hh signaling driven by genetic mutations plays a clear role in a subset of cancers.

5.3.2 Ligand-Dependent Activation

More recently, malignant activation of the Hh pathway has been attributed to high constitutive expression of Hh ligand, thereby activating the Hh pathway in the absence of activating mutations. In ligand-dependent cancers, such as chondrosarcoma and osteosarcoma, Hh ligand binds to Ptc on the tumor cell, relieving Ptc-mediated inhibition of Smo. This mode of ligand-dependent signaling may also occur in progenitor cells in certain cancers, including chronic myelogenous leukemia (CML), chronic lymphocytic leukemia (CLL), B-acute lymphocytic leukemia (B-ALL), non-Hodgkin's lymphoma (NHL), multiple myeloma (MM), and small cell lung cancer (SCLC), and may be responsible for minimal residual disease following treatment with chemotherapeutic and targeted agents (Table 5.1 and references therein). Elevated Hh ligand expression is also found in many other cancers, including pancreatic, prostate, breast, stomach, colon, hepatocellular, ovarian, medulloblastoma, and glioma (Table 5.1 and references therein). Studies of the mechanism by which elevated Hh signaling impacts these tumor types have focused on three themes that are most likely interdependent: (1) the role of direct Hh signaling to tumor cells, (2) the role of the Hh pathway in maintenance of a drug-resistant, highly metastatic tumor progenitor cell, or cancer stem cell population, and (3) the role of Hh signaling in tumor-stromal cell interactions (i.e., the tumor microenvironment).

Table 5.1 Role for the Hh pathway in solid tumors and hematologic malignancies

	Preclinical rationale	References
Solid tumors		
Colon cancer	Ligand expression, anti-tumor activity with Smo antagonism and ligand inhibition	Yauch et al. (2008), Varnat et al. (2009)
Pancreatic cancer	Ligand expression, anti-tumor activity with Smo antagonism and ligand inhibition	Thayer et al. (2003), Feldmann et al. (2008), Olive et al. (2009)
Esophageal cancer	Ligand expression	Thayer et al. (2003)
Gastric cancer	Ligand expression	Berman et al. (2003)
Hepatocellular carcinoma	Ligand expression	Sicklick et al. (2006)
Prostate cancer	Ligand expression, anti-tumor activity with Smo antagonist	Karhkadkar et al. (2004), Fan et al. (2004)
Ovarian cancer	Ligand expression, anti-tumor activity with Smo antagonism	Bhattacharya et al. (2008), Growden et al. (2009)
Breast cancer	Ligand expression, tumor-promoting activity with Hh pathway activation	Kubo et al. (2004), Liu et al. (2006), Moraes et al. (2007)
Nonsmall cell lung cancer	Ligand expression	Yuan et al. (2007)

(continued)

Table 5.1 (continued)

	Preclinical rationale	References
Small cell lung cancer	Ligand expression, tumor-promoting activity with Hh pathway activation, anti-tumor activity with Smo antagonism	Watkins et al. (2003), Vestergaard et al. (2006), Travaglione et al. (2008)
Basal cell carcinoma	Ptc or Smo mutations lead to constitutive Hh pathway activation and tumor-igenesis; anti-tumor activity with Smo antagonism	Bale and Yu (2001), Williams et al. (2003)
Melanoma	Ligand expression; Smo antagonism reduces Shh-induced vascularity	Geng et al. (2007), Stecca et al. (2007)
Chondrosarcoma	Ligand expression; Smo antagonism inhibits tumor growth	Tiet et al. (2006), Campbell et al. (2011)
Glioblastoma	Ligand expression, anti-tumor activity with Smo antagonism	Bar et al. (2007), Clement et al. (2007), Sarangi et al. (2009)
Medulloblastoma	Ptc or Smo mutations lead to constitutive Hh pathway activation and tumori-genesis; anti-tumor activity with Smo antagonism	Berman et al. (2002), Romer et al. (2004), Tremblay et al. (2009b), Robarge et al. (2009)
Hematological malignancies		
Multiple myeloma	Tumor and stromal ligand expression, inhibition of clonogenic growth with Smo antagonism or ligand inhibition	Peacock et al. (2007), Dierks et al. (2007)
Chronic mylogenous leukemia	Ligand expression, anti-tumor activity with Smo antagonism	Dierks et al. (2008), Zhao et al. (2009)
Chronic lymphocytic leukemia	Stromal ligand expression, inhibition of tumor cell survival with Smo antagonism	Hegde et al. (2008)
Acute lymphoblastic leukemia	Ligand, Ptc, Smo, Gli1 expression, self-renewal inhibited by Smo antagonists	Lin et al. (2010)
Non-Hodgkin's lymphoma	Stromal ligand expression; inhibition of clonogenic growth with Smo antagonism	Dierks et al. (2007)

5.4   Rationale for Targeting Ligand-Dependent Hh Signaling to the Tumor Cell

Hedgehog pathway plays a significant role in the biology of chondrocytes during development and in malignant tumors of the cartilage called chondrosarcoma. The Hh pathway is essential for maintaining the growth plate and trabecular bone

(Maeda et al. 2006) and for regulating chondrocyte proliferation and terminal differentiation in the endoskeleton and postembryonic growth plate (Long et al. 2001; Farquharson et al. 2001). The role of Hh in the normal biology of the chondrocyte suggested that chondrosarcoma might be a potential target for Hh pathway inhibition. Tiet et al. (2006) found evidence for constitutive Hh pathway signaling in chondrosarcomas that express high levels of Hh-regulated genes Ptc and Gli1. Chondrosarcoma tumor cell proliferation is increased by Hh ligand and decreased by inhibitors of the Hh pathway (Tiet et al. 2006). Hh pathway inhibition leads to tumor growth inhibition in 1° chondrosarcoma tumor xenografts and blocks autonomous Hh signaling in chondrosarcoma tumor cells, leading to changes in tumor morphology including loss of cellularity and increased chondroid matrix and calcification (Campbell et al. 2011). Taken together, these data provide a rationale for the evaluation of Hh inhibition in patients with chondrosarcoma.

5.5 Rationale for Targeting Hh Signaling in Minimal Residual Disease

An increasing number of cancers are being recognized as diseases that may be maintained by a biologically distinct, drug-resistant, radiation-resistant, self-renewing progenitor cell population, often called "cancer stem cells" (Clarke et al. 2006). Cancer stem cells in blood, mammary gland, gastrointestinal tract, lung, and brain cancers can share self-renewal properties and many characteristics of normal stem cells, including the expression of specific cell surface markers, expression of drug efflux pumps, and expression of genes involved in developmental pathways, including members of the Hh pathway (Clarke et al. 2006). Early evidence for tumorigenic cancer stem cells was provided in hematological malignancies, with the first report in acute myelogenous leukemia (AML) (Lapidot et al. 1994). While more difficult to prove, analogous cancer stem cells are proposed to also exist in solid tumors and contribute to tumor relapse in the setting of minimal residual disease.

In both solid tumors and hematologic malignancies, including breast cancer (Liu et al. 2006), multiple myeloma (MM) (Peacock et al. 2007), chronic myelogenous leukemia (CML) (Dierks et al. 2008; Zhao et al. 2009), acute lymphocytic leukemia (ALL) (Lin et al. 2010), and glioblastoma (Bar et al. 2007; Clement et al. 2007), these putative cancer stem cells show evidence of Hh pathway activation. Furthermore, proliferation of these cells isolated from breast cancer, glioblastoma, MM, and CML can be stimulated by the addition of Shh ligand and blocked by loss of functional Smo or with the Smo antagonist cyclopamine.

5.5.1 Hh and Cancer Stem Cells in CML

Evidence for cancer stem cells and a role for the Hh pathway are best illustrated in CML (Dierks et al. 2008; Zhao et al. 2009; Pérez-Caro et al. 2009). In a murine model of CML where BCR-ABL expression is initiated in the stem cell compartment, the resulting leukemic mice do not respond well to the Bcr-Abl inhibitor imatinib, indicating the presence of a drug-resistant subpopulation (Pérez-Caro et al. 2009). In other murine models of BCR-ABL-induced leukemia, a hematopoietic stem cell subset of the leukemic cells has evidence of elevated Hh pathway expression, with high levels of Ptc, Smo, and Gli1 (Dierks et al. 2008). In colony-forming assays using leukemic cells derived from BCR-ABL-induced leukemic mice, imatinib has minimal effects on growth of Smo-positive cells, but Smo inhibition with cyclopamine or loss of Smo expression in these cells induces apoptosis and inhibits colony formation in these assays. In vivo, loss of Smo inhibits expansion of BCR-ABL-transduced stem cells, and inhibition of Smo with cyclopamine enhances survival of BCR-ABL leukemic mice beyond that seen with the Bcr-Abl inhibitor imatinib alone. A third study (Zhao et al. 2009) confirmed these observations by showing that directed loss of Smo in the stem cell compartment decreased the incidence and severity of BCR-ABL-induced leukemia due to a reduction in CML stem cells. Conversely, when activation of Smo was directed to the stem cell compartment, BCR-ABL-induced leukemia was accelerated.

Analysis of patient CML cells has revealed a link between these findings in murine models of CML and human disease. Elevated Hh pathway expression is found in the CD34+ subset of CML cells isolated from patients (Dierks et al. 2008). Furthermore, these cells are resistant to imatinib, but sensitive to the effects of cyclopamine in colony-forming assays (Dierks et al. 2008; Zhao et al. 2009). Taken together, these studies provide persuasive evidence that CML stem cells are dependent upon Hh pathway signaling for their survival. Therefore, if Smo inhibition could eradicate CML stem cells in patients, the potential to cure this leukemia could become a reality (Quintás-Cardama et al. 2009).

5.5.2   Hh Signaling and Acute Lymphocytic Leukemia

The Hh pathway has been implicated in B-cell malignancies including MM, CLL, and NHL (Sect. 5.5.3) as well as in normal early B-cell development, suggesting that the pathway might be important in precursor B-ALL. Lin et al. (2010) showed that Hh pathway components are expressed in a variety of human precursor B-ALL cell lines and in patient-derived primary B-ALL cells that extend across a variety of cytogenetic and prognostic subgroups. In the B-ALL cell lines, basal activity of a Gli-dependent reporter is further stimulated by addition of exogenous Hh ligand and inhibited by an antibody to Hh ligand, 5E1, or the Smo inhibitors cyclopamine and IPI-926 (see below), providing evidence for autologous signaling to the tumor cell. Inhibition of Hh pathway activity in B-ALL cells mainly impacted a subpopulation of highly clonogenic B-ALL cells expressing aldehyde dehydrogenase (ALDH) and limited their self-renewal properties in vitro and their tumorigenic properties in vivo. These data demonstrate that Hh pathway activation is widespread in B-ALL and may represent a novel therapeutic target for persistent malignant clones during postremission therapy following induction.

5.5.3   Chemoresistant Progenitor Cells and SCLC

SCLC is yet another example of a tumor type that is thought to develop as a result of extensive Hh pathway activation in progenitor cells during repair of airway injury; constitutive Hh signaling has been shown to drive aberrant growth and subsequent tumor development (Watkins et al. 2003). Following tumor debulking in a primary human SCLC xenograft model with a chemotherapy treatment similar to that used clinically in patients with SCLC, administration of a Hh inhibitor leads to a significant delay in tumor regrowth (Travaglione et al. 2008). Overall, these findings suggest that the chemo-resistant, Hh-dependent "progenitor cells" may be responsible for the high relapse rate seen in SCLC patients following chemotherapy (Rudin et al. 2008). Thus, as exemplified in CML, ALL, and SCLC, the Hh pathway may be required for maintenance of the cancer stem cell population in an increasing number of cancers, and inhibition with a Hh pathway antagonist could prolong responses to conventional therapies.

5.5.4   Hh Modulation of Epithelial-Mesenchymal Transitions in Cancer

A related process that could also reflect the role of Hh in cancer stem cells is the Epithelial-to-Mesenchymal transition (EMT). EMT is a process by which cells undergo a morphological switch from a polarized epithelial phenotype to a mesenchymal/fibroblast-like phenotype and is considered an important event during tumor progression and metastasis (reviewed in Kalluri and Weinberg 2009). Signaling through the Hh pathway may cross-talk with signaling pathways that directly regulate EMT through FGF, Notch, and TGFβ signaling cascades, as well as through miRNA regulatory networks (reviewed in Katoh and Katoh 2008). The importance of EMT in tumorigenesis has been best illustrated in breast cancer. In murine models of breast cancer, cancer stem cells that arise from mammary epithelium show evidence of EMT, including fibroblastoid morphology, and increased expression of fibronectin, vimentin, and N-cadherin (Mani et al. 2008; McCoy et al. 2009; Yu et al. 2009b). In addition, induction of EMT in transformed murine mammary epithelial cells results in enrichment for cells with cancer stem cell properties, including increased tumorigenicity and acquisition of cell surface markers (Mani et al. 2008). These cells show multiple attributes of mesenchymal differentiation including expression of vimentin, fibronectin, and have increased invasive and migratory traits that can enhance the metastatic potential of the cells. A link between these observations in murine models of transformation and human disease has recently been made. Carcinoma cells on the invasive edge of tumors and those found in residual disease following conventional therapies have also been observed to undergo EMT (Creighton et al. 2009), most likely under the influence of signals originating from the surrounding stroma (see below the role of Hh in tumor-stromal interactions).

5.6   Rationale for Targeting Hh Signaling in the Tumor Microenvironment

Paracrine signaling between tumor cells and the surrounding stroma is an emerging theme for how the Hh pathway may impact many cancers, including prostate (Fan et al. 2004), pancreatic cancer (Yacht et al. 2008; Bailey et al. 2008, 2009; Olive et al. 2009), and lymphoma (Dierks et al. 2007; Hegde et al. 2008).

5.6.1   Prostate Cancer

The first evidence for Hh ligand-dependent paracrine signaling in a tumor setting was in prostate cancer. In the normal prostate, Hh signaling regulates organogenesis, but Shh ligand is also highly expressed in prostate cancer, with increased expression observed in advanced disease and in metastatic lesions (Karhkadkar et al. 2004). Studies using the LNCaP prostate tumor xenograft model have found that when Shh ligand is overexpressed by the tumor cells, Gli1 expression is activated in the surrounding tumor stroma and its activation is accompanied by enhanced tumor growth (Fan et al. 2004). Subsequent studies designed to identify the responsible cellular component(s) in the stroma found that activation of Hh signaling confined specifically to myofibroblasts is sufficient to stimulate growth of prostate tumors (Shaw et al. 2009). A gene expression signature specific for Hh-stimulated myofibroblasts was identified and shown to be associated with the presence of reactive stroma in a subset of patient-derived prostate tumors. This gene expression signature correlated with production of Hh ligand by the tumor and Gli1 upregulation in the surrounding stroma (Shaw et al. 2009). These studies demonstrate that Hh signaling from tumor cells to the stroma could elicit tumor growth-promoting effects that originate in the stroma.

5.6.2   Pancreatic Cancer

Hh ligand-dependent paracrine signaling has also been well documented in pancreatic cancer (Yacht et al. 2008; Bailey et al. 2008, 2009). One of the hallmarks of pancreatic cancer is marked proliferation of stromal fibroblasts and deposition of a dense extracellular matrix, a phenomenon known as "desmoplasia." Desmoplasia is believed to contribute to tumor progression by creating protumorigenic effects in the tumor microenvironment (Mahadevan and Von Hoff 2007). Investigations into why pancreatic cancer is one of the more chemo-resistant tumor types have found that extensive desmoplasia is associated with an abnormal vasculature that provides a barrier to drug delivery due to poor perfusion of the tumors (Olive et al. 2009). A high percentage of pancreatic tumors express elevated levels of Hh ligand and, like in prostate cancer, tumor-derived Hh ligand leads to activation of Gli1 in stromal myofibroblasts (Yacht et al. 2008; Bailey et al. 2008). Hh ligand-dependent paracrine signaling in pancreatic cancer can stimulate differentiation, migration, and proliferation of myofibroblasts, which can promote tumor-associated desmoplasia (Bailey et al. 2008). A murine model of pancreatic cancer, called KPC (mutant Kras and p53), closely resembles many aspects of the human disease, including tumor progression, histopathological appearance with desmoplasia, elevated expression of Shh ligand, poor tumor perfusion, and a lack of responsiveness to the chemotherapeutic agent gemcitabine (Olive et al. 2009). In an effort to test whether inhibition of Hh signaling would improve drug delivery to tumors and the survival of KPC mice, a small molecule inhibitor, IPI-926 (see below), was administered orally to KPC mice, alone and in combination with gemcitabine. IPI-926, which inhibits the Hh pathway by binding to Smo, resulted in down-modulation of Gli1 mRNA in the pancreatic tumor tissue, indicating inhibition of Hh pathway signaling. In concordance with the ascribed role of Hh signaling in promoting desmoplasia, histopathological analysis revealed that treatment with IPI-926 resulted in decreased stromal content due to decreased proliferation of myofibroblasts. The "hypostromal" appearance was accompanied by increased vascularity and improved delivery of multiple agents to the tumors, including gemcitabine. The combined treatment with IPI-926 and gemcitabine led to increased apoptosis of tumor cells, decreased metastases to the liver, and a doubling of the median survival time of tumor-bearing animals when compared to the gemcitabine-alone treated mice. Many clinical investigations have tried to improve upon the activity of gemcitabine in pancreatic cancer with minimal success. These findings in the KPC mouse model suggest that disruption of Hh ligand-dependent paracrine signaling may reduce or eliminate desmoplasia and enhance the delivery and activity of therapeutics in patients with pancreatic tumors.

5.6.3 Hh Signaling Between B-Cell-Derived Hematologic Malignancies and the Microenvironment

B-cell-derived hematologic malignancies are known to be highly dependent upon the stromal microenvironment for growth and survival (Kurtova et al. 2009). Recent studies have provided convincing evidence that Hh ligand produced by stromal cells plays a role in growth and survival of B-cell-derived malignancies including NHL, MM, and CLL (Dierks et al. 2007; Hegde et al. 2008). Hh ligands (Shh and Ihh) secreted by bone marrow and lymph node stromal cells function as survival factors for tumor cells including malignant lymphoma and plasmacytoma cells that are derived from transgenic Eµ-Myc mice (Dierks et al. 2007). Inhibition of hedgehog signaling in vivo induced apoptosis through down-regulation of Bcl-2 and prevented the expansion of Eµ-Myc murine lymphoma cells in syngeneic mice and reduced tumor mass in mice with fully developed disease. Immunohistochemistry showed that, in contrast to what has been observed in prostate and pancreatic cancer, Hh ligands are produced only by stromal cells, whereas Smo and Gli1 expression were detected in lymphoma tumor cells. Notably, when NHL and MM tumor cells were isolated from patients and cocultured with a stromal cell line that produced Ihh, survival of the tumor cells was inhibited by cyclopamine or a blocking antibody to Hh ligand. These studies provided the first evidence that stromally induced Hh signaling may provide an important survival signal for B-cell malignancies in vitro and in vivo.

Further studies with patient-derived B-CLL cells have found that CLL cells do not survive in vitro unless bone marrow or lymph node-derived stromal cells are present (Kurtova et al. 2009). In a coculture system, Hh ligands were found to be expressed in stroma from bone marrow and lymph nodes, and these ligand-expressing stromal cells prevented B-CLL cells from undergoing apoptosis (Hegde et al. 2008). Hh-ligand-dependent activation of Gli1 expression in B-CLL cells is inhibited by the Smo antagonist, cyclopamine. Inhibition of Hh signaling using either cyclopamine or anti-sense oligonucleotides specific for Gli1 sensitized B-CLL cells to the chemotherapy, fludarabine. Notably, B-CLL cells from patients with poor clinical outcome have higher expression of Gli1 when compared to better clinical outcome subgroups. In total, these findings suggest that Hh ligand-dependent signaling is active between the stroma and several types of B-cell-derived tumor cells and that Hh inhibition could be a relevant strategy for improving response to chemotherapy.

In summary, Hh ligand-dependent paracrine signaling plays an important role in how tumor cells and the surrounding stromal microenvironment interact to promote tumor growth, regulate tumor cell differentiation, impart drug resistance, prevent drug delivery, and provide an appropriate niche for cancer stem cells. Thus, while the outcome of paracrine signaling between a tumor and the stromal microenvironment may differ depending upon the tumor-specific context, there is compelling evidence that inhibition of Hh ligand-dependent paracrine signaling could be a viable therapeutic approach to complement existing chemotherapies and targeted agents.

5.7   Clinical Evaluation of Hh Pathway Inhibition in Cancer: Smoothened Antagonists in Clinical Trials

The role of Smo in malignant activation of the Hh pathway in such a wide range of cancers represents an opportunity for Smo antagonists to have a substantial impact as cancer therapeutics. Above, numerous examples have been provided to show how Smo inhibition can target the cancer cell as well as the tumor microenvironment. Importantly, while the Hh pathway orchestrates numerous processes throughout embryogenesis and development, it appears largely inactive or dispensable in most adult tissues. Thus, Hh pathway inhibition has the potential to be selective against cancer and provides a manageable side-effect profile. This rationale has compelled multiple biopharmaceutical companies to invest considerable resources into the development of Hh pathway inhibitors. At the writing of this review, eight new chemical entities have entered clinical trials and are in various stages of testing (Table 5.2). To date, all Hh pathway inhibitors under evaluation in the clinic are orally administered small molecule inhibitors that target Smo. Chemical structures of IPI-926, GDC-0449, LDE225, PF04449913, and TAK-441 are published (Tremblay et al. 2009a, b; Robarge et al. 2009; Pan et al. 2010; Jackson-Fisher et al. 2011; Tojo et al. 2011).

Table 5.2 Smo antagonists in clinical trials

Company	Mechanism of inhibition	Trial status[a]
Genentech/Curis (GDC-0449)	Oral Smo antagonist	Ph 1, Ph 1b, Ph 2
		BCC, BCNS, adult and pediatric medulloblastoma, colon, ovarian, SCLC, glioblastoma, sarcoma
BMS/Exilexis (BMS-833923)	Oral Smo antagonist	Ph 1 solid tumors; Ph 1b
		BCC, multiple myeloma, gastric, esophageal, SCLC; Ph 1/2 CML
Infinity Pharmaceuticals, Inc. (IPI-926)	Oral Smo antagonist	Ph 1, BCC and solid tumors
		Ph 1 HNSCC
		Ph 1b/2 pancreatic cancer
		Ph 2 chondrosarcoma
Novartis (LDE225)	Oral and topical Smo antagonist	Ph 1 solid tumors, adult and pediatric medulloblastoma
		Ph 2 BCC, BCNS
Novartis (LEQ506)	Smo antagonist	Ph 1
		Medulloblastoma, BCC, and solid tumors
Pfizer (PF04449913)	Oral Smo antagonist	Ph 1
		Hematological malignancies, CML
Lilly (LY2940680)	Oral Smo antagonist	Ph 1 solid tumors
Takeda-Millenium (TAK-441)	Oral Smo antagonist	Ph 1 solid tumors

[a] http//www.clinicaltrials.gov

5.7.1 IPI-926

IPI-926 is a novel semisynthetic derivative of the natural product cyclopamine that directly binds to and blocks the activity of Smo. IPI-926 was designed to improve upon the chemical stability, solubility, potency, selectivity, oral bioavailability, and metabolic stability of the natural product (Tremblay et al. 2009b). In nonclinical studies conducted in four species, IPI-926 has high oral bioavailability (50–100%), a long plasma half-life (8 to >24 h), and a high volume of distribution (9–30 L/kg). It is highly selective for Smo and inhibits Hh pathway activation in cell-based assays with EC50s of <10 nM. IPI-926 has demonstrated biological activity in multiple preclinical animal models of cancer, including Hh ligand-independent medulloblastoma (Tremblay et al. 2009b; Pink et al. 2008; Villavicencio et al. 2009), where daily oral administration leads to dose-dependent inhibition of tumor growth and complete tumor regression at higher doses. In Hh ligand-dependent tumor types, such as ovarian cancer and SCLC, IPI-926 administration leads to a significant delay in tumor regrowth following debulking with chemotherapy (Growden et al. 2009; Travaglione et al. 2008). Furthermore, as described above, IPI-926 treatment of mice bearing pancreatic cancer results in Hh pathway inhibition in the tumor-associated stroma, depletion of desmoplasia, and improved drug delivery, resulting in prolonged survival of the mice when administered in combination with gemcitabine

(Olive et al. 2009). In primary chondrosarcoma tumor xenograft models, IPI-926 inhibited tumor growth and downregulated expression of Gli1 and Ptc in the tumor cells (Campbell et al. 2011).

IPI-926 was investigated in a Phase 1 clinical trial in patients with advanced and/ or metastatic solid tumor malignancies. In this trial, IPI-926 was well tolerated up to 160 mg/day with the most common adverse events being fatigue, nausea, and elevated transaminases that were asymptomatic and reversible (Rudin et al. 2011). Inhibition of Gli1 expression was observed in normal skin and tumor biopsies from BCC patients and evidence of clinical activity was observed in a cohort of BCC patients. IPI-926 is being evaluated in a randomized Phase 1b/2 trial in combination with gemcitabine in patients with previously untreated metastatic pancreatic cancer with overall survival as the endpoint. In the Phase 1b portion of this study, the combination of daily IPI-926 plus standard doses of gemcitabine was well tolerated with observed adverse events that were consistent with the known safety profile of each agent (Stephenson et al. 2011). Thirty-one percent of patients had a partial response. The randomized Phase 2 trial is ongoing with 160 mg IPI-926 administered daily with the standard dose of gemcitabine. A second randomized Phase 2 trial is ongoing in patients with metastatic or locally advanced (unresectable) chondrosarcoma (Table 5.2).

5.7.2 GDC-0449

The Smo antagonist GDC-0449 is a synthetic small molecule discovered through a high-throughput Hh-dependent cell-based screen (Robarge et al. 2009). The resulting benzimidazole series was optimized for potency, PK, and drug-like properties by heterocyclic replacements and further modifications to the amide portion of the molecule resulting in GDC-0449. In preclinical studies, GDC-0449 produced dose-dependent, complete tumor regression with twice-daily oral administration in a medulloblastoma mouse model (Robarge et al. 2009). In preliminary results from the first Phase 1 study of GDC-0449, anti-tumor activity was observed in cancers where malignant activation of Smo is ligand-independent and driven by activating mutations in Ptc. In 33 locally advanced or metastatic BCC patients, the overall response rate was 55% and the drug was well tolerated (Von Hoff et al. 2009).

In addition to responses in BCC patients in the Phase 1 trial, an adult patient with medulloblastoma also showed a dramatic response, but within 2 months developed resistance to GDC-0449 (Rudin et al. 2009). In this patient, resistance to treatment was attributed to a single amino acid mutation in Smo that was similarly demonstrated in medulloblastoma-bearing mice while undergoing treatment with GDC-0449 (Yacht et al. 2009). GDC-0449 is under evaluation in multiple clinical trials including three company-sponsored trials that have completed enrollment: (1) a pivotal Phase 2 trial for locally advanced or metastatic BCC, (2) in combination with chemotherapy and bevacizumab as front-line therapy in patients with colorectal cancer, and (3) as maintenance therapy in patients with ovarian cancer after a second or third complete response (Table 5.2). Recently, the Phase 2 colorectal trial failed to meet its primary endpoint of improvement in progression-free survival with the addition of GDC-0449 to chemotherapy and bevacizumab (Berlin et al. 2010). Furthermore, analysis of data from the Phase 2 ovarian trial indicates that, although GDC-0449 treatment did not significantly prolong progression-free survival in this clinical study, there was a positive trend in improved progression-free survival in ovarian cancer patients receiving GDC-0449 following a second complete response (Kaye et al. 2010). Trials in additional cancers are ongoing in cooperation with the National Cancer Institute and are actively enrolling patients with BCC medulloblastoma, pancreatic cancer, SCLC, gastric and esophageal cancer, glioblastoma, and others (Table 5.2).

Results from the first Phase 1 trial have led to the conduct of a trial in Gorlin's syndrome patients (Epstein et al. 2011) to determine the safety and efficacy of GDC-0449 in BCC in this patient population. Interim results from the trial in Gorlin's patients showed that GDC-0449 treatment prevents development of new BCC lesions and reduces the size of existing lesions. Gli1 expression was reduced and decreased proliferation, as assessed by Ki67 staining, was observed in the BCC lesions. The anti-BCC activity in Gorlin's patients was accompanied by side effects including muscle cramps, hair loss, and loss of taste, causing some patients to discontinue drug wherein the side effects were resolved. Due to the highly significant differences in efficacy between the placebo and GDC-0449 treated groups at the interim analysis, the placebo arm of the trial was terminated early.

5.7.3 NVP LDE225

NVP LDE225 is derived from a novel biphenyl-3-carboxamide chemical series identified via high-throughput screening and optimized for Smo antagonism (Pan et al. 2010). Administration of LDE225 to mice-bearing subcutaneous Ptch/−p53−/− medulloblastoma allografts led to dose-related tumor growth inhibition with tumor regression observed in the higher dosing groups. This compound is currently in Phase 1 clinical trials for evaluation of clinical PK, efficacy, and safety (Table 5.2). In the Phase 1 trial, LDE225 was well tolerated up to 800 mg/day with dose-limiting adverse events consisting of elevated plasma creatine phosphokinase associated with myalgia (Tawbi et al. 2011). Inhibition of Gli1 expression was observed in skin and tumor biopsies and anti-tumor activity was seen in patients with medulloblastoma and basal cell carcinoma. Multiple Phase 2 trials are ongoing in patients with basal cell carcinoma and a Phase 1 trial has been initiated in children with recurrent or refractory medulloblastoma and other tumors potentially dependent on Hh signaling.

5.8 Conclusions

The results described above, while exciting for Phase 1 studies, highlight important questions for the future of drugs that inhibit the Hh pathway. Will resistance that develops to one Smo antagonist impact sensitivity to Smo antagonists from different structural classes? Results obtained thus far suggest that availability of multiple Smo antagonists could benefit patients by having additional inhibitors on hand to treat tumors that have developed resistance-conferring mutations, similar to the development of multiple tyrosine kinase inhibitors that have differential activity against various mutations found in a given kinase (Baselga 2006). In addition, will other Hh ligand-independent tumors, like BCC, be subject to development of resistance-conferring mutations?

The larger clinical opportunity for Hh inhibitors is in the setting of Hh ligand-dependent signaling between the tumor and the microenvironment. However, the negative Phase 2 trials with GDC-0449 point out the challenge of determining where ligand-dependent signaling is relevant in the setting of a chemoresponsive tumor types like colorectal cancer or ovarian cancer. Given that genetic instability in the stroma would not be expected, optimism remains high that development of resistance-conferring mutations in that setting will be less likely. With clinical validation of Hh inhibition achieved in the Hh ligand-independent setting, the challenge remains to determine where and how inhibitors will be best employed in cancers with Hh ligand-dependent signaling. Whether single agent activity will be observed in the maintenance setting of other chemo-responsive tumors will be further addressed in ongoing trials, e.g., SCLC. Likewise, ongoing combination studies in pancreatic cancer will examine whether Hh inhibition can positively impact the activity of chemotherapy and targeted agents in tumor types that are much less responsive to standard-of-care chemotherapies. Clearly, with an expanding inventory of cancers that demonstrate activation of the Hh pathway and an increased understanding of how the pathway impacts specific tumor types, inhibitors of Hh signaling could be invaluable additions to the arsenal of new therapies to target cancer.

References

Alcedo J, Ayzenzon M, Von Ohlen T, Noll M, Hooper JE (1996) The *Drosophila* smoothened gene encodes a seven-pass membrane protein, a putative receptor for the hedgehog signal. Cell 86:221–232

Bailey JM, Swanson BJ, Hamada T, Eggers JP, Singh PK, Caffery T, Ouellette MM, Hollingsworth MA (2008) Sonic hedgehog promotes desmoplasia in pancreatic cancer. Clin Cancer Res 14:5995–6004

Bailey JM, Mohr AM, Hollingsworth MA (2009) Sonic hedgehog paracrine signaling regulates metastasis and lymphangiogenesis in pancreatic cancer. Oncogene 28:3513–3525

Bale AE, Yu KP (2001) The hedgehog pathway and basal cell carcinomas. Hum Mol Genet 10:757–762

Bar EE, Chaudhry A, Lin A, Fan X, Schreck K, Matsui W, Piccirillo S, Vescovi AL, DiMeco F, Olivi A (2007) Cyclopamine-mediated hedgehog pathway inhibition depletes stem-like cancer cells in glioblastoma. Stem Cells 25:2524–2533

Baselga J (2006) Targeting tyrosine kinases in cancer: the second wave. Science 312:1175–1178

Belloni E, Muenke M, Roessler E, Traverso G, Siegel-Bartelt J, Frumkin A, Mitchell HF, Donis-Keller H, Helms C, Hing AV, Heng HH, Koop B, Martindale D, Rommens JM, Tsui LC, Scherer SW (1996) Identification of sonic hedgehog as a candidate gene responsible for holoprosencephaly. Nat Genet 14:353–356

Berlin J, Bendell J, Hart LL, Firdaus I, Gore I, Hermann RC, Mackey H, Yacht B, Graham RA, Low JA, Bray GL (2010) Phase II study of GDC-0449 (HPI) with concurrent chemotherapy and bevacizumab as first-line therapy for metastatic colorectal cancer. In: Proceedings of the European Society of Medical Oncology, Milan, Italy, 8–12 Oct 2010, Abstract #LBA21

Berman DM, Karhadkar SS, Hallahan AR, Pritchard JI, Eberhart CG, Watkins DN, Chen JK, Cooper MK, Taipale J, Olson JM, Beachy PA (2002) Medulloblastoma growth inhibition by hedgehog pathway blockade. Science 297:1559–1561

Berman DM, Karhadkar SS, Maitra A, Montes de Oca R, Gerstenblith MR, Briggs K, Parker AR, Shimada Y, Eshleman JR, Watkins DN, Beachy PA (2003) Widespread requirement for hedgehog ligand stimulation in growth of digestive tract tumors. Nature 425:846–851

Bhattacharya R, Kwon J, Ali B, Wang E, Patra S, Shridhar V, Mukherjee P (2008) Role of hedgehog signaling in ovarian cancer. Clin Cancer Res 14:7659–7666

Campbell VT, Nadesan P, Wang Y, Whetstone H, McGovern K, Read MA, Alman BA, Wunder JS (2011) Direct targeting of the Hedgehog pathway in primary chondrosarcoma xenografts with the smoothened inhibitor IPI-926. In: Proceedings of the American Association of Cancer Research Annual Meeting, Orlando, FL, 2–6 Apr 2011, Abstract LB380

Chen JK, Taipale J, Cooper MK, Beachy PA (2002) Inhibition of hedgehog signaling by direct binding of cyclopamine to smoothened. Genes Dev 16:2743–2748

Chen W, Ren XR, Nelson CD, Barak LS, Chen JK, Beachy PA, de Sauvage F, Lefkowitz RJ (2004) Activity-dependent internalization of smoothened mediated by beta-arrestin 2 and GRK2. Science 306:2257–2260

Chiang C, Litingtung Y, Lee E, Young KE, Corden JL, Westphal H, Beachy PA (1996) Cyclopia and defective axial patterning in mice lacking sonic hedgehog gene function. Nature 383:407–413

Clarke MF, Dick JE, Dirks PB, Eaves CJ, Jamieson CHM, Jones DL, Visvader J, Weissman IL, Wahl GM (2006) Cancer stem cells – perspective on current status and future directions: AACR workshop on cancer stem cells. Cancer Res 66:9339–9344

Clement V, Sanchez P, de Tribolet N, Radovanovic I, Ruiz i Altaba A (2007) Hedgehog-Gli1 signaling regulates human glioma growth, cancer stem cell self-renewal, and tumorigenicity. Curr Biol 17:165–172

Cooper MK, Porter JA, Young KE, Beachy PA (1998) Teratogen-mediated inhibition of target tissue response to *Shh* signaling. Science 280:1603–1607

Creighton CJ, Lia X, Landis M, Dixon JM, Neumeister VM, Sjolund A, Rimm DL, Wong H, Rodriguez A, Herschkowitz JI, Fan C, Zhang X, He X, Pavlick A, Gutierrez MC, Renshaw L, Larionov AA, Faratian D, Hilsenbeck SG, Perou CM, Lewis MT, Rosen JN, Chang JC (2009) Residual breast cancers after conventional therapy display mesenchymal as well as tumor-initiating features. Proc Natl Acad Sci U S A 106:13820–13825

Das S, Harris LG, Metge BJ, Liu S, Riker AI, Samant RS, Shevde LA (2009) The hedgehog pathway transcription factor GLI1 promotes malignant behavior of cancer cells by up-regulating osteopontin. J Biol Chem 284:22888–22897

Dierks C, Grbic J, Zirlik K, Beigi R, Englund NP, Guo G-R, Veelken H, Engelhardt M, Mertelsmann R, Kelleher JF, Schultz P, Warmuth M (2007) Essential role of stromally induced hedgehog signaling in B-cell malignancies. Nat Med 13:944–951

Dierks C, Beigi R, Guo G-R, Zirlik K, Stegert MR, Manley P, Trussell C, Schmitt-Graeff A, Landwerlin K, Veelken H, Warmuth M (2008) Expansion of Bcr-Abl-positive leukemic stem cells is dependent on hedgehog pathway activation. Cancer Cell 14:238–249

Eichberger T, Kaser A, Pixner C, Schmid C, Klingler S, Winklmayr M, Hauser-Kronberger C, Aberger F, Frischauf AM (2008) GLI2-specific transcriptional activation of the bone morphogenetic protein/activin antagonist follistatin in human epidermal cells. J Biol Chem 283:12426–12437

Epstein E, Tang JY, Mackay-Wiggan JM, Aszterbaum M, Lindgren J, Chang K, Coppola C, Campbell A, Chanana A, Marji J, Callahan C, Yacht R, Bickers DR (2011) An investigator-initiated, phase II randomized, double-blind, placebo-controlled trial of GDC-0449 for prevention of BCCs in basal cell nevus syndrome (BCNS) patients. In: Proceedings of the American Association of Cancer Research, Orlando, FL, 2–6 Apr 2011, Abstract #LB01

Fan L, Pepicelli CV, Dibble CC, Catbagan W, Zarycki JL, Laciak R, Gipp J, Shaw A, Lamm MLG, Munoz A, Lipinski R, Thrasher JB, Bushman W (2004) Hedgehog signaling promotes prostate xenograft tumor growth. Endocrinology 145:3961–3970

Farquharson C, Jefferies D, Seawright E, Houston B (2001) Regulation of chondrocyte terminal differentiation in the postembryonic growth plate: the role of the PtHrP-Indian Hh axis. Endocrinology 142:4131–4140

Feldmann G, Habbe N, Dhara S, Bisht S, Alvarez H, Fendrich V, Beaty R, Mullendore M, Karikari C, Bardeesy N, Ouellette MM, Yu W, Maitra A (2008) Hedgehog inhibition prolongs survival in a genetically engineered mouse model of pancreatic cancer. Gut 57:1420–1430

Geng L, Cuneo KC, Cooper MK, Wang H, Sekhar K, Fu A, Hallahan DF (2007) Hedgehog signaling in the murine melanoma microenvironment. Angiogenesis 10:259–267

Goodrich LV, Milenkovi HKM, Scott MP (1997) Altered neural cell fates and medulloblastoma in mouse *patched* mutants. Science 277:1109–1113

Growden WB, McCann CR, Curley M, Friel AM, Mandley E, Ferguson J, Foster R, MacDougal J, Rueda BR (2009) Hedgehog pathway inhibitor cyclopamine suppresses Gli1 expression and inhibits serous ovarian cancer xenograft growth (Abstract). In: Society of Gynecologic Oncologists annual meeting, San Antonio, TX, 5–8 Feb 2009, Abstract 18

Hahn H, Wicking C, Zaphiropoulos PG, Gailani MR, Shanley S, Chidambaram A, Vorechovsky I, Holmberg E, Unden AB, Gillies S, Negus K, Smyth I, Pressman C, Leffell DJ, Gerrard B, Goldstein AM, Dean M, Toftgard R, Chenevix-Trench G, Wainwright B, Bale AE (1996) Mutations of the human homolog of *Drosophila* patched in the nevoid basal cell carcinoma syndrome. Cell 85:841–851

Han Y-G, Kim HJ, Dlugosz AA, Ellison DW, Gilbertson RJ, Alvarez-Buylla AA (2009) Dual and opposing roles of primary cilia in medulloblastoma development. Nat Med 15:1062–1065

He J, Sheng T, Stelter AA, Li C, Xhang X, Sinha M, Luxon BA, Xie J (2006) Suppressing Wnt signaling by the hedgehog pathway through sFRP-1. J Biol Chem 281:35598–35602

Hegde GV, Peterson KJ, Emanuel K et al (2008) Hedgehog-induced survival of B-cell chronic lymphocytic leukemia cells in a stromal cell microenvironment: a potential new therapeutic target. Mol Cancer Res 6:1928–1936

Incardona JP, Gaffield W, Kapur RP, Roellink H (1998) The teratogenic ceratrum alkaloid cyclopamine inhibits sonic hedgehog signal transduction. Development 125:3553–3562

Jackson-Fisher AJ, McMahon MJ, Lam J, Li C, Engstrom LD, Tsaparikos K, Shields DJ, Fang DD, Lira ME, Zhu Z, Robbins MD, Schwab R, Munchhof MJ, VanArsdale T (2011) PF-04449913, a small molecule inhibitor of Hedgehog signaling, is effective in inhibiting tumor growth in preclinical models. In: Proceedings of the American Association of Cancer Research Annual Meeting, Orlando, FL, 2–6 Apr 2011, Abstract #4504

Jiang J, Hui C (2008) Hedgehog signaling in development and cancer. Dev Cell 15:801–812

Kalluri R, Weinberg RA (2009) The basics of epithelial-mesenchymal transition. J Clin Invest 119:1420–1428

Karhkadkar SS, Bova GS, Abdallah N, Dhara S, Gardner D, Maltra A, Isaccs JT, Berman DM, Beachy PA (2004) Hedgehog signalling in prostate regeneration, neoplasia and metastasis. Nature 431:707–712

Katoh Y, Katoh M (2008) Hedgehog signaling: epithelial-to-mesenchymal transition and miRNA (review). Int J Mol Med 22:271–275

Kaye S, Fehrenbacher L, Holloway R, Horowitz N, Karlan B, Amit A, Slomovitz B, Chang I, Yauch RL, Reddy JC (2010) Phase 2 randomized placebo controlled study of hedgehog pathway inhibitor GDC-0449 as maintenance therapy in patients with ovarian cancer in 2nd or 3rd complete remission. In: Proceedings of the European Society of Medical Oncology, Milan, Italy, 8–12 Oct 2010, Abstract #LBA25

Keeler RF (1968) Teratogenic compounds of *Veratrum californicum* (Durand)-IV. Phytochemistry 7:303–306

Kovacs JJ, Whalen EJ, Liu R, Xiao K, Kim J, Chen M, Wang J, Chen W, Lefkowitz RJ (2008) β-Arrestin mediated localization of smoothened to the primary cilium. Science 320:1777–1781

Kubo M, Nakamura M, Tasaki A, Yamanaka N, Nakashima H, Nomura M, Kuroki S, Katano M (2004) Hedgehog signaling pathway is a new target for patients with breast cancer. Cancer Res 64:6071–6074

Kurtova AV, Balakrishnan K, Chen R, Ding W, Schnabl S, Quiroga MP, Sivina M, Wierda WG, Estrov Z, Keating MJ, Shehata M, Jäger U, Gandhi V, Kay NE, Plunkett W, Burger JA (2009) Diverse marrow stromal cells protect CLL cells from spontaneous and drug-induced apoptosis: development of a reliable and reproducible system to assess stromal cell adhesion-mediated drug resistance. Blood 114:4441–4450

Lapidot T, Sirard C, Vormoor J, Murdoch B, Hoang T, Caceres-Cortes J, Minden M, Paterson B, Caligiuri MA, Dick JE (1994) A cell initiating human acute myeloid leukaemia after transplantation into SCID mice. Nature 367:645–648

Li X, Deng W, Nail CD, Bailey SK, Kraus MH, Ruppert JM, Lobo-Ruppert SM (2006) Snail induction is an early response to Gli1 that determines the efficiency of epithelial transformation. Oncogene 25:609–621

Lin TL, Wang QH, Brown P, Peacock C, Merchant AA, Brennan S, Jones E, McGovern K, Watkins DN, Sakamoto KM, Matsui W (2010) Self-renewal of acute lymphocytic leukemia cells is limited by the hedgehog pathway inhibitors cyclopamine and IPI-926. PLoS One 5:e15262

Liu A, Wang B, Niswander LA (2005) Mouse intraflagellar transport proteins regulate both the activator and repressor functions of Gli transcription factors. Development 132:3103–3111

Liu S, Dontu G, Mantle ID, Patel S, Ahn N, Jackson KW, Suri P, Wicha MS (2006) Hedgehog signaling and Bmi-1 regulate self-renewal of normal and malignant human mammary stem cells. Cancer Res 66:6063–6071

Long F, Zhang XM, Karp S, Yang Y, McMahon AP (2001) Genetic manipulation of Hh signaling in the endochondral skeleton reveals a direct role in the regulation of chondrocyte proliferation. Development 128:5099–5108

Maeda Y, Nakamura E, Nguyen M-T et al (2006) Indian Hh produced by postnatal chondrocytes is essential for maintaining a growth plate and trabecular bone. Proc Natl Acad Sci U S A 104:6382–6387

Mahadevan D, Von Hoff DD (2007) Tumor stroma interactions in pancreatic ductal adenocarcinoma. Mol Cancer Ther 6:1186–1197

Mani SA, Guo W, Liao MJ, Eaton EN, Ayyanan A, Zhou AY, Brooks M, Reinhard F, Zhang CC, Shipitsin M, Campbell LL, Polyak K, Brisken C, Yang J, Weinberg RA (2008) The epithelial-mesenchymal transition generates cells with properties of stem cells. Cell 133:704–715

McCoy EL, Iwanaga R, Jedlicka P, Abbey NS, Chodosh LA, Heichman KA, Welm AL, Ford HL (2009) Six1 expands the mouse mammary epithelial stem/progenitor cell pool and induces mammary tumors that undergo epithelial-mesenchymal transition. J Clin Invest 119:2528–2531

Mill P, Mo R, Hu MC, Dagnino L, Rosenblum ND, Hui CC (2005) Shh controls epithelial proliferation via independent pathways that converge on N-Myc. Dev Cell 9:293–303

Moraes RC, Zhang X, Harrington N, Fung JY, Wu MF, Hilsenbeck SG, Allred DC, Lewis MT (2007) Constitutive activation of smoothened (SMO) in mammary glands of transgenic mice leads to increased proliferation, altered differentiation, and ductal dysplasia. Development 134:1231–1242

Nüsslein-Volhard C, Wieschaus E (1980) Mutations affecting segment number and polarity in *Drosophila*. Nature 287:795–801

Ogden SK, Fei DL, Schilling NS, Ahmed YF, Hwa J, Robbins DJ (2008) G protein Gα_i functions immediately downstream of smoothened in hedgehog signalling. Nature 456(7224):967–970

Olive KP, Michael J, Davidson CJ, Gopinathan A, McIntyre D, Honess D, Madhu B, Goldgraben MA, Caldwell ME, Allard D, Frese KK, DeNicola G, Feig C, Combs C, Winter SP, Ireland H, Reichelt S, Howat WJ, Chang A, Dhara M, Wang L, Rückert F, Grützmann R, Pilarsky C, Izeradjene K, Hingorani SR, Huang P, Davies SE, Plunkett W, Egorin M, Hruban RH, Whitebread N, McGovern K, Adams J, Iacobuzio-Donahue C, Griffiths J, Tuveson DA (2009) Inhibition of hedgehog signaling enhances delivery of chemotherapy in a mouse model of pancreatic cancer. Science 324:1457–1461

Oliver TG, Grasfeder LL, Carroll AL, Kaiser C, Gillingham CL, Lin SM, Wickramasinghe R, Scott MP, Wechsler-Reya RJ (2003) Transcriptional profiling of the sonic hedgehog response: a critical role for N-myc in proliferation of neuronal precursors. Proc Natl Acad Sci U S A 100:7331–7336

Pan S, Wu X, Jiqing J, Gao W, Wan Y, Cheng D, Huan D, Liu J, Englund JP, Wang Y, Peukert S, Miller-Moslin K, Yuan J, Guo R, Matsumoto M, Vattay A, Jiang Y, Tsao J, Sun F, Pferdekamper AC, Dodd S, Tuntland T, Maniara W, Kelleker JF, Yao Y, Warmuth M, Williams J, Dorsch M (2010) Discovery of NVP-LDE225, a potent and selective smoothened antagonist. Med Chem Lett 1:130–134

Peacock CD, Wang Q, Gesell GS, Cocoran-Schwartz IM, Jones E, Kim J, Devereux WL, Rhodes JT, Huff CA, Beachy PA, Watkins DN, Matsui W (2007) Hedgehog signaling maintains a tumor stem cell compartment in multiple myeloma. Proc Natl Acad Sci U S A 104:4048 4053

Pérez-Caro M, Cobaleda C, González-Herrero I, Vicente-Dueñas C, Bermejo-Rodríguez C, Sánchez-Beato M, Orfao A, Pintado B, Flores T, Sánchez-Martín M, Jiménez R, Piris MA, Sánchez-García I (2009) Cancer induction by restriction of oncogene expression to the stem cell compartment. EMBO J 28:8–20

Pink M, Proctor J, Briggs KJ, MacDougall J, Whitebread N, Tremblay MR, Grogan M, Palombella VJ, Castro AC, Adams J, Read MA, Corcoron-Schwartz IM, Harcke T, Eberhart CG, Watkins DN, Sydor J (2008) Activity of IPI-926, a potent HH pathway inhibitor, in a novel model of medulloblastoma derived from Ptch/HIC +/- mice (Abstract). In: Proceedings of the 99th Meeting of the American Association for Cancer Research, San Diego, CA, 12–16 Apr 2008, Abstract 1588

Quintás-Cardama A, Kantarjian H, Cortes J (2009) Imatinib and beyond – exploring the full potential of targeted therapy for CML. Nat Rev Clin Oncol 6:535–543

Regl G, Kasper M, Schnidar H, Eichberger T, Neill GW, Philpott MP, Esterbauer H, Hauser-Kronberger C, Frischauf AM, Aberger F (2004) Activation of the BCL2 promoter in response to Hedgehog/GLI signal transduction is predominantly mediated by GLI2. Cancer Res 64:7724–7731

Robarge KD, Brunton SA, Castanedo GM, Cui Y, Dina MS, Goldsmith R, Gould SE, Guichert O, Gunzner JL, Halladay J, Jia W, Khojasteh C, Koehler MF, Kotkow K, La H, Lalonde RL, Lau K, Lee L, Marshall D, Marsters JC Jr, Murray LJ, Qian C, Rubin LL, Salphati L, Stanley MS, Stibbard JH, Sutherlin DP, Ubhayaker S, Wang S, Wong S, Xie M (2009) GDC-0449-a potent inhibitor of the hedgehog pathway. Bioorg Med Chem Lett 19:5576–5581

Roessler E, Belloni E, Gaudenz K, Jay P, Berta P, Scherer SW, Tsui LC, Muenke M (1996) Mutations in the human sonic hedgehog gene cause holoprosencephaly. Nat Genet 14:357–360

Rohatgi R, Milenkovic L, Scott MP (2007) Patched-1 regulates hedgehog signaling at the primary cilium. Science 317:372–376

Rohatgi R, Milenkovic L, Corcoran RB, Scott MP (2009) Hedgehog signal transduction by smoothened: pharmacologic evidence for a 2-step activation process. Proc Natl Acad Sci U S A 106:3196–3201

Romer JT, Kimura H, Magdaleno S, Sasai K, Fuller C, Baines H, Connelly M, Stewart CF, Gould S, Rubin LL, Curran T (2004) Suppression of the Shh pathway using a small molecule inhibitor eliminated medulloblastoma in *Ptc1+/-p53-/-* mice. Cancer Cell 6:229–240

Rudin CM, Hann CL, Peacock CD, Watkins DN (2008) Novel systemic therapies for small cell lung cancer. J Natl Compr Canc Netw 6(3):1–8

Rudin CM, Hann CL, Laterra J, Yauch RL, Callahan CA, Fu L, Holcomb T, Stinson J, Gould SE, Coleman B, LoRusso PM, Von Hoff DD, de Sauvage FJ, Low JA (2009) Treatment of medulloblastoma with hedgehog pathway inhibitor GDC-0449. N Engl J Med 361:1173–1178

Rudin CM, Jimeno A, Miller WH, Eigl B, Gettinger S, Chang A, Faia K, Sweeney J, Loewen G, Ross R, Weiss GJ (2011) A phase 2 study of IPI-926, a novel hedgehog pathway inhibitor, in patients with advance or metastatic solid tumors. In: Proceedings of the American Society of Clinical Oncology, Chicago, IL, 2–7 Jun 2011, Abstract #3014

Sarangi A, Valadez JG, Rush S, Abel TW, Thompson RC, Cooper MK (2009) Targeted inhibition of the hedgehog pathway in established malignant glioma xenografts enhances survival. Oncogene 28:3468–3476

Sasai K, Romer JR, Lee Y, Finkelstein D, Fuller C, McKinnon PJ, Curran T (2006) Shh pathway activity is down-regulated in cultured medulloblastoma cells: implications for preclinical studies. Cancer Res 66:4215–4222

Shaw A, Gipp J, Bushman W (2009) The sonic hedgehog pathway stimulates prostate tumor growth by paracrine signaling and recapitulates embryonic gene expression in tumor myofibroblasts. Oncogene 28:4480–4490

Sicklick JK, Li Y-X, Jayaraman A, Kannangai R, Qi Y, Vivekanandan P, Ludlow JW, Owzar K, Chen W, Torbenson MS, Diehl AM (2006) Dysregulation of the hedgehog pathway in human hepatocarcinogenesis. Carcinogenesis 27:748–757

Stecca B, Mas C, Clement V, Zbinden M, Correa R, Piguet V, Beermann F, Ruiz I, Altaba A (2007) Melanomas require HEDGEHOG-GLI signaling regulated by interactions between GLI1 and the RAS-MEK/AKT pathways. Proc Natl Acad Sci U S A 104:5895–5900

Stephenson J, Richards D, Wolpin B, Becerra C, Hamm J, Messersmith W, Devens S, Cushing J, Goddard J, Schmalbach T, Fuchs CS (2011) The safety of IPI-926, a novel hedgehog pathway inhibitor, in combination with gemcitabine in patients with metastatic pancreatic cancer. In: Proceedings of the American Society of Clinical Oncology, Chicago, IL, 2–7 Jun 2011, Abstract #4114

Taipale J, Chen JK, Cooper MK, Wang B, Mann RK, Milenkovic L, Scott MP, Beachy PA (2000) Effects of oncogenic mutations in smoothened and patched can be reversed by cyclopamine. Nature 406:1005–1009

Tawbi HA, Ahnert JR, Dummer R, Thomas AL, Granvil C, Shou Y, Dey J, Mita MM, Amakye DD, Mita AC (2011) Phase I study of LDE225 in advanced solid tumors: updated analysis of safety, preliminary efficacy, and pharmacokinetic-pharmacodynamic correlation. In: Proceedings of the American Society of Clinical Oncology, Chicago, IL, 2–7 Jun 2011, Abstract #3062

Thayer SP, di Magliano MP, Heiser PW, Nielsen CM, Roberts DJ, Lauwers GY, Qi YP, Gysin S, Fernández-del Castillo C, Yajnik V, Antoniu B, McMahon M, Warshaw A, Hebrok M (2003) Hedgehog is an early and late mediator of pancreatic cancer tumorigenesis. Nature 425:851–856

Tiet TD, Hopyan S, Nadesan P, Gokgoz N, Poon R, Lin AC, Yan T, Andrulis IL, Alman BA, Wunder JS (2006) Constitutive hedgehog signaling in chondrosarcoma up-regulates tumor cell proliferation. Am J Pathol 168:321–330

Tojo H, Shibata S, Satoh Y, Kawamura M, Inazuka M, Yamakawa H, Kashiwag M, Miyamoto M, Kondo M, Oohashi T, Oguro Y, Sasaki S (2011) TAK-441, a novel investigational small molecule hedgehog pathway inhibitor for use in cancer therapy. In: Proceedings of the American Association of Cancer Research Annual Meeting, Orlando, FL, 2–6 Apr 2011, Abstract 4508

Travaglione V, Peacock CD, MacDougall J, McGovern J, Cushing J, Yu LC, Trudeau M, Palombella V, Adams J, Hierman J, Rhodes J, Devereux WL, Watkins DN (2008) A novel HH pathway inhibitor, IPI-926, delays tumor recurrence post-chemotherapy in a primary human SCLC xenograft model (Abstract). In: Proceedings of the 99th meeting of the American Association for Cancer Research, San Diego, CA, 12–16 Apr 2008, Abstract 4611

Tremblay MR, Nesler M, Weatherhead R, Castro AC (2009a) Recent patents for hedgehog pathway inhibitors for the treatment of malignancy. Expert Opin Ther Pat 19:1039–1056

Tremblay M, Lescarbeau A, Grogan MJ, Tan E, Lin G, Austad BC, Yu LC, Behnke ML, Nair SJ, Hagel M, White K, Conley J, Manna JD, Alvarez-Diez TM, Hoyt J, Woodward CN, Sydor JR, Pink M, MacDougall J, Campbell MJ, Cushing J, Ferguson J, Curtis MS, McGovern K, Read MA, Palombella VJ, Adams J, Castro AC (2009b) Discovery of a potent and orally active hedgehog pathway antagonist (IPI-926). J Med Chem 52:4400–4418

Varnat F, Duquet A, Malerba M, Zbinden M, Mas C, Gervaz P, I Altaba AR (2009) Human colon cancer epithelial cells harbor active hedgehog-Gli signaling that is essential for tumour growth, recurrence, metastasis and stem cell survival and expansion. EMBO Mol Med 1:1–14

Vestergaard J, Pedersen MW, Pedersen N, Ensinger C, Tümer Z, Tommerup N, Poulsen HS, Larsen LA (2006) Hedgehog signaling in small-cell lung cancer: frequent in vivo but a rare event in vitro. Lung Cancer 52:281–290

Villavicencio E, Khanna P, Ditzler S, Pullar B, Hansen S, Read MA, Faia K, Hatton B, Knoblaugh S, Randolph-Habecker J, LeBlanc M, Shaw D, Friedman S, McGovern K, Olson J (2009)

Activity of the Shh pathway inhibitor, IPI-926, in a mouse model of medulloblastoma. In: Proceedings of the 100th meeting of the American Association of Cancer Research, Denver, CO, 18–22 Apr 2009, Abstract #3199

Von Hoff DD, LoRusso PM, Rudin CM, Reddy JC, Yauch RL, Tibes R, Weiss GJ, Borad MJ, Hann CL, Brahmer JR, Mackey HM, Lum BL, Darbonne WC, Marsters JC Jr, de Sauvage FJ, Low JA (2009) Inhibition of the hedgehog pathway in advanced basal-cell carcinoma. N Engl J Med 361:1164–1172

Watkins DN, Berman DM, Burkholder SG, Wang B, Beachy PA, Baylin SB (2003) Hedgehog signaling within airway epithelial progenitors and in small-cell lung cancer. Nature 422:313–317

Welch KD, Panter KE, Lee ST, Gardner DR, Stegelmeier BL, Cook D (2009) Cyclopamine-induced synophthalmia in sheep: defining a critical window and toxicokinetic evaluation. J Appl Toxicol 29:414–421

Williams JA, Guicherit OM, Zaharian BI, Xu Y, Chai L, Wichterle H, Kon C, Gatchalian C, Porter JA, Rubin LL, Wang FY (2003) Identification of a small molecule inhibitor of the hedgehog signaling pathway: effects on basal cell carcinoma-like lesions. Proc Natl Acad Sci U S A 100:4616–4621

Wong SY, Seol AD, So P-L, Ermilov AN, Bichakjian CK, Epstein EH, Dlugosz AA, Reiter JR (2009) Primary cilia can both mediate and suppress hedgehog pathway-dependent tumorigenesis. Nat Med 15:1055–1061

Xie J, Johnson RL, Zhang X, Bare JW, Waldman FM, Cogen PH, Menon AG, Warren RS, Chen L-C, Scott MP, Epstein EH Jr (1997) Mutations of the *patched* gene in several types of sporadic extracutaneous tumors. Cancer Res 57:2369–2372

Yauch RL, Gould SE, Scales SJ, Tang T, Tian H, Ahn CP, Marshall D, Fu L, Januario T, Kallop D et al (2008) A paracrine requirement for hedgehog signaling in cancer. Nature 455:406–410

Yauch RL, Dijkgraaf GJ, Alicke B, Januario T, Ahn CP, Holcomb T, Pujara K, Stinson J, Callahan CA, Tang T, Bazan JF, Kan Z, Seshagiri S, Hann CL, Gould SE, Low JA, Rudin CM, de Sauvage FJ (2009) Smoothened mutation confers resistance to a hedgehog pathway inhibitor in medulloblastoma science. Science 326(5952):572–574

Yoon JW, Kita Y, Frank DJ, Majewski RR, Konicek BA, Nobrega MA, Jacob H, Walterhouse D, Iannaccone P (2002) Gene expression profiling leads to identification of GLI1-binding elements in target genes and a role for multiple downstream pathways in GLI1-induced cell transformation. J Biol Chem 277:5548–5555

Yu M, Gipp J, Yoon JW, Iannaccone P, Walterhouse D, Bushman W (2009a) Sonic hedgehog-responsive genes in the fetal prostate. J Biol Chem 287:5620–5629

Yu M, Smolen GA, Zhang J, Wittner B, Schott BJ, Brachtel E, Ramaswamy S, Maheswaran S, Haber DA (2009b) A developmentally regulated inducer of EMT, LBX1, contributes to breast cancer progression. Genes Dev 23:1737–1742

Yuan Z, Goetz JA, Singh S, Ogden SK, Petty WJ, Black CC, Memoli VA, Dmitrovsky E, Robbins DJ (2007) Frequent requirement of hedgehog signaling in non-small cell lung carcinoma. Oncogene 26:1046–1055

Zhao C, Chen A, Jamieson CH, Fereshteh M, Abrahamsson A, Blum J, Kwon HY, Kim J, Chute JP, Rizzieri D, Munchhof M, VanArsdale T, Beachy PA, Reya T (2009) Hedgehog signalling is essential for maintenance of cancer stem cells in myeloid leukaemia. Nature 458:776–779

Chapter 6
Wnt Signaling in Cancer Pathogenesis and Therapeutics

Naoko Takebe, Lawrence Lum, and S. Percy Ivy

6.1 Introduction

Cancer-initiating cells frequently exploit mechanisms that promote self-renewal in stem cells to sustain tumor growth. Cellular programs controlled by secreted signaling molecules such as the Wnt, Hedgehog (Hh), Notch/Delta, fibroblast growth factor (FGF), and bone morphogenic proteins (BMPs), initially identified as regulators of cell fate determination in embryonic development, have more recently been shown to promote growth in cancerous cells. This review will focus on the Wnt signal transduction pathway which appears to be deregulated in a majority of colorectal cancers and in a variety of other cancers as well (de Lau et al. 2007). In the majority of cancer incidents, loss of negative regulator function, either by epigenetic silencing or loss-of-function mutations, gives rise to aberrant Wnt pathway responses. Despite the strong genetically based rationale for targeting the Wnt pathway in many of these cancers, the lack of suitable targets within the pathway has limited our ability to test the utility of this molecularly targeted therapeutic approach. This review will provide an overview of current progress in targeting the dysregulated Wnt signaling pathway in cancer.

6.1.1 Wnt Signaling Pathway

The Wnt family of secreted proteins directs cell fate determination at various stages of development and in homeostatic contexts (Logan and Nusse 2004; Reya and Clevers 2005). The human Wnt gene family consists of 19 highly conserved

N. Takebe • L. Lum • S.P. Ivy (✉)
Investigational Branch, Cancer Therapy Evaluation Program,
National Cancer Institute, NIH, Rockville, MD 20852, USA
e-mail: Ivyp@ctep.nci.nih.gov

D.A. Frank (ed.), *Signaling Pathways in Cancer Pathogenesis and Therapy*,
DOI 10.1007/978-1-4614-1216-8_6, © Springer Science+Business Media, LLC 2012

cysteine-rich glycoproteins (Nusse 2005). The term Wnt is derived from the fusion of two gene names: "Wingless (Wg)", the first Wnt gene identified in *Drosophila* from a genetic screen for genes that regulate embryonic patterning (Sharma 1973; Rijsewijk et al. 1987), and "Int-1", a murine proto-oncogene initially identified as a gene that induces mammary tumor formation and later identified as a Wnt protein (Nusse and Varmus 1982; Nusse et al. 1991). Although much of our understanding of Wnt-mediated signaling is derived from studies of embryonic development, an abundance of evidence supports a role for this gene family in regeneration of many adult tissues, including bone and the intestinal epithelium.

Wnt proteins are categorized with respect to the cellular response they elicit upon binding to the transmembrane receptor Frizzled (Fzd). Those Wnt proteins that elicit a β-catenin-dependent transcriptional response are termed "canonical" Wnt proteins, whereas those that elicit other responses are termed "non-canonical." The noncanonical Wnt pathway employs a broad range of cytoplasmic signaling molecules and is generally less well understood than the canonical response (Chien et al. 2009). Designating Wnt signaling pathways as either canonical or noncanonical is in many cases overly simplistic. Indeed, given that the mammalian genome encodes for 19 Wnt proteins and 10 Fzd receptors, potentially 190 Wnt/Fzd pairing combinations exist.

6.1.2 Wnt Protein Production

The Wnt proteins are typically acylated with a palmitoyl adduct most frequently added by the Porcupine (Porcn) extracellular acyltransferase (Kadowaki et al. 1996; Nusse 2003) (Fig. 6.1). Fatty acylation of Wnt proteins is likely essential to the normal biosynthesis of these molecules (van den Heuvel et al. 1993; Couso and Martinez Arias 1994; Willert et al. 2003). Transport of Wnt proteins through the secretory pathway also requires the action of another protein known as Wntless/Evi, a multi-pass transmembrane protein that likely functions as a chaperone molecule (Banziger et al. 2006; Bartscherer et al. 2006). The intercellular transport of Wnt molecules is likely mediated by glypican receptors, heparan sulfate proteoglycans that bind to a broad range of secreted molecules (Baeg and Perrimon 2000; Cadigan 2008).

6.1.3 The Wnt/β-Catenin Signaling Cascade

The Wnt/β-catenin signaling cascade is initiated when specific Wnt isoforms bind to their corresponding transmembrane receptors encoded by the *Frizzled* gene family, together with a member of the LDL receptor protein (LRP5 or 6 (van Amerongen et al. 2008) (Fig. 6.2). The Wnt-Fzd interaction leads to recruitment of a large cyto-plasmic regulatory complex scaffolded by the Axin and APC tumor suppressor

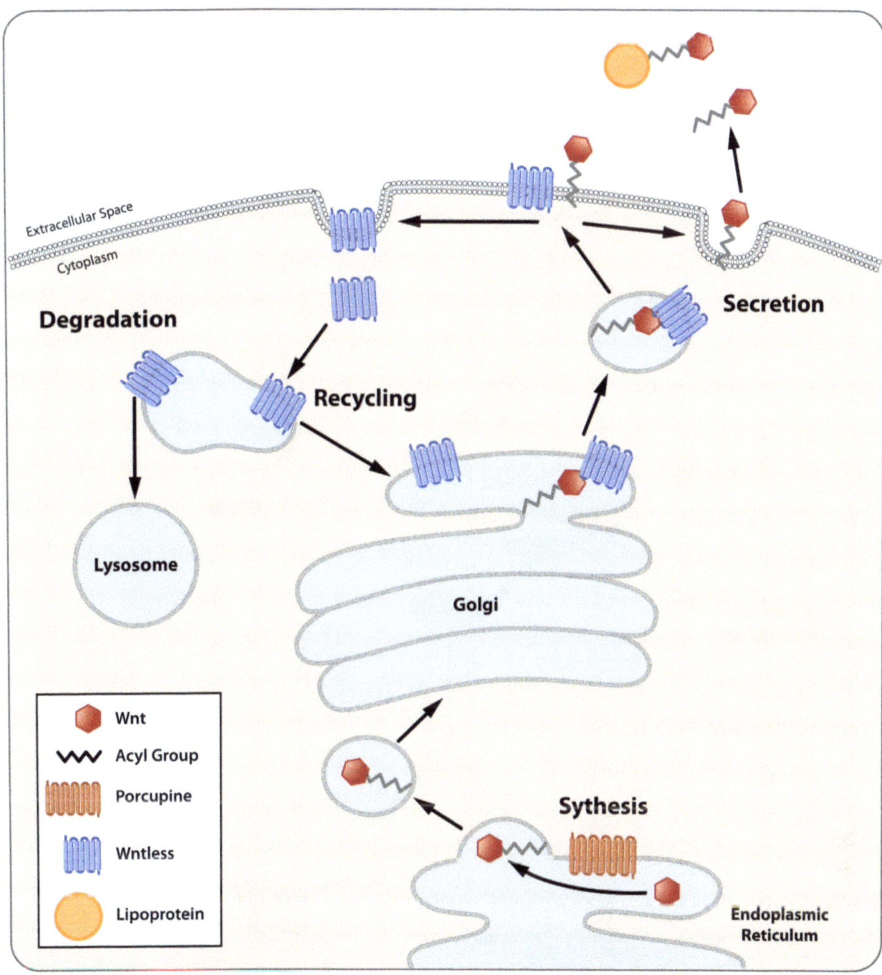

Fig. 6.1 Wnt Protein Production The Wnt protein is palmitoylated by the Porc acyltransferase. Wntless guides Wnt to the cell surface from Golgi to interact with glypicans or lipoproteins within the extracellular space. Wintless is recycled back to the Golgi or degraded by the lysosomes (Adapted from Cadigan 2008)

proteins to the membrane. The activity of glycogen synthase kinase 3beta (GSK)-3β, a serine/threonine kinase that phosphorylates and marks β-catenin for proteasome-mediated destruction in the absence of Wnt signaling, is inhibited. The β-catenin that accumulates is able to bind to members of the T-cell factor/lymphoid enhancer factor (Tcf/Lef) family of DNA-binding proteins. As Axin proteins also scaffold GSK3β and β-catenin, these scaffolding molecules serve dual functions as both activators and suppressors of Wnt/β-catenin pathway responses. Indeed, Axin proteins are also required for LRP phosphorylation (Jiang et al. 2008).

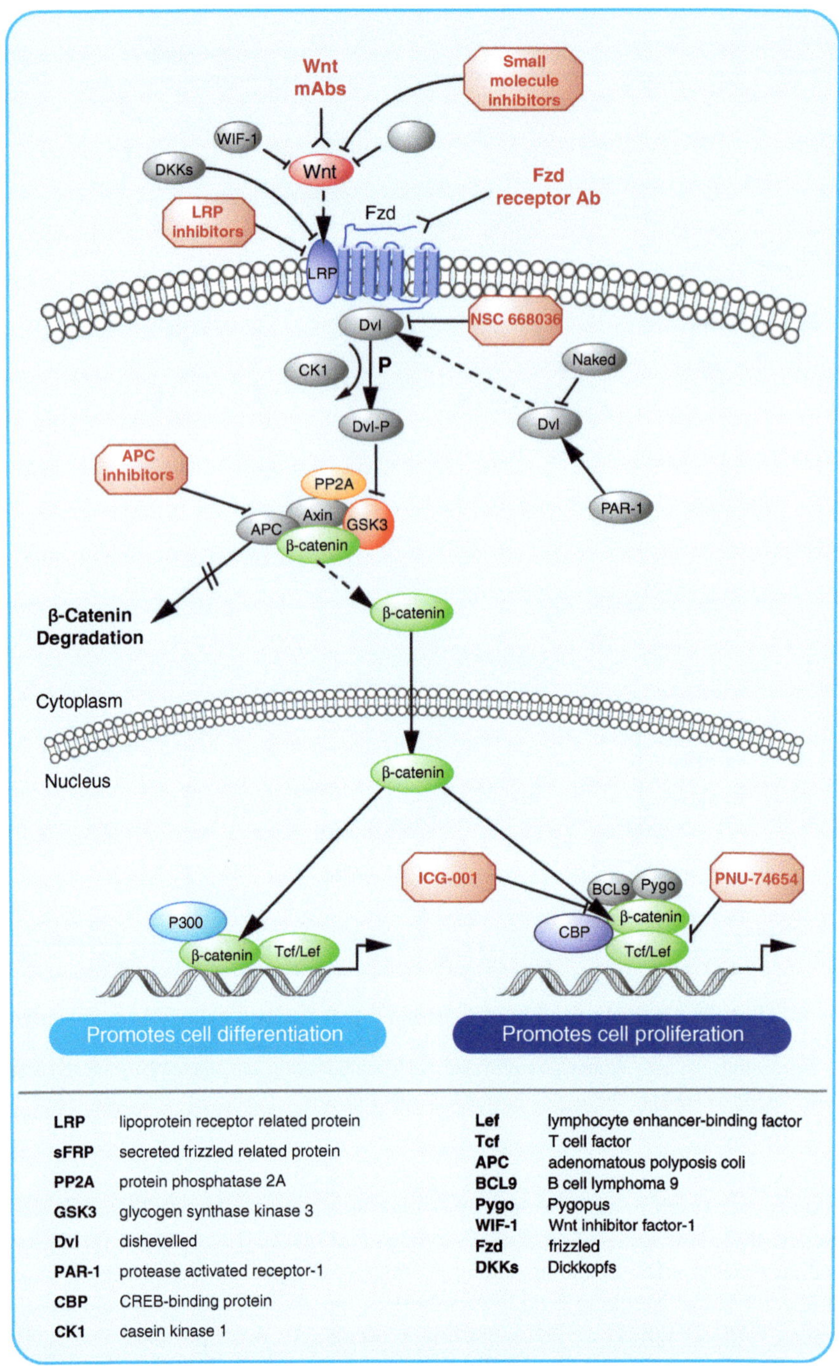

Fig. 6.2 Wnt/β-Catenin Signaling Cascade and Investigational Therapeutics Wnt ligand binding is inhibited by SFRPs and WIF-1. DKKs inhibit the LRP directly. Pharmacologic interventions

Regulatory mechanisms at the transcription level additionally play a role in tuning Wnt pathway responses. The interaction of β-catenin and Tcf/Lef molecules is also regulated by proteins that include TLE/Gro, ICAT, Chibby, Sox9, or CtBP-APC factor. Some of these interactions, such as those between TCF and TLE/Gro molecules, confer repressor function to TCF/LEF molecules (Brantjes et al. 2001). Adding greater complexity to the action of Tcf/Lef proteins are the numerous forms of each protein that can be generated as a result of alternative splicing (Arce et al. 2006; Tang et al. 2008). Activation of the Wnt/β-catenin pathway then entails β-catenin-dependent displacement of the repressor molecules that bind to Tcf/Lef proteins in a splice-variant-dependent manner by activation. In vertebrate systems, β-catenin and cyclic AMP-responsive element-binding protein (CREB)-binding protein (CBP)/p300 have been shown to interact directly (Hecht et al. 2000; Miyagishi et al. 2000; Sun et al. 2000; Takemaru and Moon 2000) and activate transcriptional targets (Emami et al. 2004; Ma et al. 2005). Wnt/β-catenin signaling has been shown to be dichotomous, serving to control both cellular proliferation and differentiation. Recently, a selective agent for CBP/β-catenin interaction (ICG-001) enabled the researchers to develop a model to explain the divergent activities of this signaling (see Table 6.1) (McMillan and Kahn 2005). At the transcriptional level, the switch from β-catenin/CBP to β-catenin/p300 may control one of the most fundamental cell switch points; CBP/β-catenin mediates transcription for stem/progenitor cell maintenance and proliferation. The switch to p300/β-catenin can be mediated pharmacologically or through naturally occurring differentiation factors (McMillan and Kahn 2005).

6.1.4 Wnt/β-Catenin Signaling and Tumorigenesis

Aberrant nuclear accumulation of β-catenin can be detected in colorectal, gastric, esophageal, melanoma, lung, ovarian, cervical, endometrial, breast, prostate, thyroid, hepatoblastoma, hepatocellular, medulloblastoma, and pancreatic carcinomas (Giles et al. 2003), suggesting a pathogenic role for Wnt/β-catenin signaling pathway. The pathway is also activated in Wilms tumor (Koesters et al. 1999), acute lymphocytic leukemia (McWhirter et al. 1999), and multiple myeloma cells (Derksen et al. 2004). Approximately 30% of hepatocellular carcinomas (Reed et al. 2008) and 90% of human colorectal carcinomas have deregulated Wnt signaling pathways (Fevr et al. 2007).

Fig. 6.2 (continued) include small molecule inhibitors and Wnt monoclonal antibodies (mAb). Fzd receptor Ab inhibits Fzd receptor directly. NSC668036 inhibits Dvl causing stabilization of the degradation complex. Inactivated destruction complex causes stabilization of β-catenin levels before nuclear translocation and binding to the DNA-binding transcription complex (includes Tcf/Lef, CBP, BCL9, and Pygo). ICG-001 inhibits β-catenin and CBP interaction resulting in inhibition of cell proliferation and differentiation

Table 6.1 Modulators of Wnt signaling

Target	Agents in development	Mechanism of action	Phase of development
Wnt ligands	• Wnt1 mAb • Wnt2 mAb • Soluble Wnt receptor Frizzled8CRD-hFc Ab (decoy receptor)	Bind to extracellular ligands resulting in apoptosis by decreased activity of the transcription factor	Preclinical
Frizzled (Fzd) receptors	• Fzd1 receptor Ab • Fzd2 receptor Ab	Inhibit ligand-receptor interaction	Preclinical
Disheveled (Dvl) family members	• NSC668036 • FJ9 (disrupts Fzd-7 and Disheveled)	Inhibition of Dvl causes stabilization of degradation complex resulting in β-catenin degradation	Preclinical
β-catenin reverse nuclear transport (Yoshizumi et al. 2004)	Thiazolidinedione (TZD) (anti-diabetic drug)	Transport β-catenin from nuclear to plasma membrane	Preclinical in cancer settings
β-catenin/TCF (van Stolk et al. 2000; Emami et al. 2004; Arber et al. 2006)	• PNU-74654 (β-catenin-Tcf) • ICG-001 (CBP-β-catenin) • NSAIDs (Cox2 inhibitor, Salmedix, etc.)	Inhibit protein–protein interaction resulting in decreased β-catenin-dependent gene expression	Preclinical (except for NSAIDS – Phase 2)
Protein degradation process (Cong et al. 2003; Su et al. 2003; Liu et al. 2004; Nagao et al. 2008; Chen et al. 2009)	• AV65 • Artificial F-box protein • Sulindac	Exact mechanism unknown. Thought to enhance proteasome degradation of β-catenin via ubiquitin-conjugated proteolysis	Preclinical (except for Sulindac – Phase 2)
Axin2 (Chen et al. 2009; Huang et al. 2009)	• IWR (inhibitors of Wnt response) • XAV939	Both compounds stabilize Axin proteins causing β-catenin loss likely by targeting tankyrase enzymes that promote Axin destruction	Preclinical
SFRP family members	Either Ab approach as above or small molecule inhibitors	Decrease ligand concentration in extracellular space	Preclinical
WIF family members		Antagonize interaction with extracellular ligands	Preclinical
Porcupine (porc) (Chen et al. 2009)	IWP (inhibitors of Wnt production)	Inhibits biosynthesis of Wnt proteins by blocking Porcn-mediated Wnt protein acylation	Preclinical

The frequency of APC mutations in cancer suggests that disengagement of normal β-catenin regulation by Wnt ligands is the predominant mechanism for cancerous Wnt/β-catenin pathway initiation. Loss of APC function is observed in both hereditary tumor syndromes (familial adenomatous polyposis or FAP) (Polakis 2000) or sporadic tumors (Segditsas et al. 2008). Indeed, increased activity of β-catenin due to mutated APC has been a key initiating factor in colorectal carcinoma (Boman and Wicha 2008). Less frequently, mutations that affect β-catenin phosphorylation and thus destruction are found in cancer with perhaps the highest occurrence rate in desmoid tumors (Tejpar et al. 1999).

Engagement of Wnt/β-catenin pathway response resulting from loss of secreted Wnt antagonists, such as Fzd-related proteins (sFRPs), Dickkopfs (DKKs) and Wnt inhibitory factor-1 (WIF1), has also been observed (Finch et al. 1997; Glinka et al. 1998; Hsieh et al. 1999).

6.1.5 Wnt Signaling and Cancer Stem Cells (CSCs)

The shared dependency of carcinogenesis and tissue homeostasis on Wnt-mediated cellular responses supports the existence of cancer stem cells that perpetually drive tumor growth through their sustained self-renewal. Indeed, deletion of the APC gene in intestinal crypt stem cells is sufficient to induce tumorigenesis (Barker et al. 2009). Additional evidence for a role of Wnt pathway signaling in supporting cancer stem cell renewal has been gleaned from studies of leukemia. Notably, β-catenin accumulation in granulocyte-macrophage progenitor cells is observed in CML blast crises and supports a mechanism for transforming committed progenitors into leukemia stem cells (Jamieson et al. 2004).

The cutaneous CSC maintenance has been shown to depend on the Wnt/β-catenin signaling pathway (Malanchi et al. 2008). Elimination of the β-catenin gene in a mouse model resulted in loss of cutaneous stem cells and complete regression of DMBA/TPA-induced tumors. The tumor regression was characterized by extensive terminal differentiation of the remaining cells. In this study, β-catenin-deficient tumor cells (devoid of CD34+ biomarker) lost their ability to initiate secondary tumors even when transplanted in numbers up to 10^6. An analysis of human squamous cell carcinomas (SCC) also indicated constitutive activation of the Wnt/β-catenin signaling pathway, as suggested by the presence of nuclear β-catenin. The dependence of human SCC on the Wnt/β-catenin signaling pathway may provide therapeutic opportunities using agents that target this pathway.

6.1.6 Wnt Signaling and Cancer Metastasis

Recently, Wnt/Tcf signaling and its target genes HOXB 9 and Lef1 have been identified as mediators of chemotactic invasion and colony outgrowth in a lung adenocarcinoma model. In this model, activation of the canonical Wnt/Tcf pathway is found as a determinant of metastasis to brain and bone during lung adenocarcinoma progression (Nguyen et al. 2009). Wnt signaling drives epithelial to mesenchymal transitions (EMTs) through inhibition of GSK3β-mediated destruction complex activity, causing increased levels of nuclear β-catenin and activation of EMT-inducing transcription factors (Vincan and Barker 2008). However, accumulation of β-catenin alone does not appear to be sufficient to induce EMT. The majority of colorectal carcinomas exhibit genetic inactivation of APC and concomitant activation of Wnt/β-catenin pathway responses, though few of these tumors exhibit any mesenchymal features (Polyak and Weinberg 2009). Thus, Wnt-mediated responses are likely to collaborate with other cell fate determination pathways in promoting EMT in cancer (Bailey et al. 2007).

6.1.7 Biomarkers of Wnt Signaling

The utility of Wnt expression levels as prognostic indicators in various types of cancers is still unknown. Loss of Wnt5a has been shown to correlate with decreased time to recurrence and decreased survival in node-negative colorectal carcinoma, invasive breast cancer, and neuroblastoma patients (Jonsson et al. 2002; Blanc et al. 2005; Dejmek et al. 2005). In contrast, in ovarian cancer, melanoma, and gastric carcinoma patients, an increased Wnt5a expression showed a correlation with poor survival (Kurayoshi et al. 2006; Da Forno et al. 2008; Badiglian Filho et al. 2009). Thus, meaningful interpretation of Wnt protein expression levels as prognostic indicators would entail a deep understanding of Wnt protein function in a tissue-dependent context. Similarly, expression levels of sFRP-2 and Dkk-3 appear to be associated with bladder cancer (Urakami et al. 2006) with expression of membranous sFRP-4 correlating with a significant improvement in relapse-free survival in this disease (Horvath et al. 2004). Clinically useful pharmacodynamic biomarker development seems to be developed most effectively when a novel agent is available in the preclinical stage. Detecting the accumulation and nuclear localization of β-catenin is a useful means of examining Wnt/β-catenin signaling activation (Groen et al. 2008).

6.2 Noncanonical Wnt Signaling Cascades and Cancer

The term "non-canonical" is applied to those Wnt molecules or Wnt-activated signaling pathways that do not engage β-catenin-induced transcription activation. The majority of noncanonical Wnt signals is initiated through FzdR family receptors and receptor tyrosine kinase-like orphan receptor 2 (ROR2)/related to receptor tyrosine kinase (RYK) coreceptors and transmitted via disheveled and its downstream effector molecules [Rho family GTPasese, c-jun and NH2-terminal kinase (JNK)] or by changes in intracellular calcium that in turn activate Nemo-like kinase (NLK) or nuclear factor of activated T cells (NFAT)-signaling cascades (Oishi et al.

2003; Lu et al. 2004). Our understanding of noncanonical Wnt signaling has been limited by the complexity of cellular responses elicited in a tissue-type dependent manner. This is further complicated by the absence of general markers for measuring pathway activity. For instance, the role of the noncanonical Wnt5a signaling in cancer likely functions as either a tumor suppressor or proto-oncogene depending on the cancer-type (McDonald and Silver 2009). Despite in vitro evidence suggesting that noncanonical Wnt pathway responses may contribute to cancerous growth, genetic evidence that supports anti-cancer effects from disrupting these responses is still lacking.

6.3 Wnt Pathway as Target for Anticancer Drug Development

6.3.1 Preclinical Agents Targeting Wnt Pathway Responses

The preponderance of cancer-associated mutations in APC has focused therapeutic developmental efforts towards the identification of small molecules that disengage Wnt/β-catenin signaling via downstream regulatory mechanisms. These include those that induce β-catenin destruction and those that perturb transcriptional activity of the TCF/LEF molecules. ICG-001 is the first selective small inhibitor of β-catenin-Tcf-mediated transcription identified by a high-throughput small-molecule library screening (Emami et al. 2004). ICG-001 disrupts interaction between β-catenin and CBP. Engagement of β-catenin with CBP- or the CBP homologs protein p300 appears to be a switch that controls CSC differentiation (McMillan and Kahn 2005). Interestingly, ICG-001 does not appear to disrupt CBP interaction with p300.

Another level of Wnt pathway response that appears to be amenable to chemical perturbation is β-catenin destruction. Recently, two new classes of small molecule compounds were identified from high-throughput synthetic chemical library screens that modulate levels of Axin proteins, which assemble the β-catenin destruction complex. In one study, the Inhibitor of Wnt response (IWR) compounds were shown to abrogate destruction of Axin proteins, thus preventing accumulation of β-catenin (Chen et al. 2009). In the context of cells with APC mutations, the stabilized Axin molecules functionally compensate for the lack of APC activity at least with respect to regulation of β-catenin protein levels. More recently, these compounds, along with another compound known as XAV939, have been shown to target the tankyrase enzymes, novel regulators of Axin protein stability (Huang et al. 2009). From the same study that identified the IWR compounds, a second class of compounds termed Inhibitor of Wnt production (IWP) revealed Porcn, an enzyme essential for Wnt protein biosynthesis. Thus, the identification of two novel targets, Porcn and tankyrase enzymes, has improved the chances of establishing therapeutic strategies premised upon regulation of the Wnt pathway. These molecules are also likely to be of value in dissecting out the roles of noncanonical and canonical signaling in a broad range of biological systems. Lastly, the novel compound AV65 appears to promote β-catenin destruction and has shown activity by inducing apoptosis in a dose-dependent manner against CML cell lines, though the molecular target is unknown (Nagao et al. 2008). Nevertheless, AV65 appears to be useful in either imatinib (ST1571, Gleevec®) or second-generation Abl TKI-resistant CML patients.

6.3.2 Clinical Investigational Agents Targeting Wnt Pathway Responses

Although their mechanism of action is less clear, a number of approved compounds for treatment of diseases not strongly associated with Wnt pathway responses have shown activity against Wnt/β-catenin pathway response. These include imatinib, originally identified as a PDGFR inhibitor (Buchdunger et al. 2000), which appears to inhibit tyrosine phosphorylation of β-catenin (Rao et al. 2006). Similarly, COX inhibitors including Celecoxib (Celebrex®) have been shown to inhibit β-catenin activity in human colon carcinoma cells by inducing its degradation (Maier et al. 2005). Lastly, thiazolidinedione (TZD), which inhibits nuclear translocation of β-catenin, has been reported to completely inhibit metastases in a colon cancer xenograft model (Yoshizumi et al. 2004). A better understanding of the specificity of these compounds for targeting Wnt-dependent processes and their mechanism of action in the Wnt pathway may identify approved compounds that could be prescribed as anti-cancer agents.

6.4 Summary/Future Directions

The discovery of therapeutic agents inhibiting aberrant Wnt-dependent signaling responses holds several challenges. First, if the primary targets of these agents are CSCs, then the standard tumor volume measurement criteria may not be an appropriate tool for the efficacy endpoint in clinical trial. Thus, new definitions of therapeutic efficacy may need to be established. Second, because of the cross-talk with other cell-signaling pathways, inhibition of Wnt pathway responses may lead to feedback effects that counteract the effect of inhibitors. Rationally designed combination regimens based on careful preclinical studies in specific indications are the most appropriate strategy for development of these agents including novel–novel combinations. Third, appropriate biomarkers are important for predicting responses and for pharmacodynamic evaluation. These should be developed at an early stage of clinical testing. And lastly, as seen with other targeted agents, new classes of agents may hold promise for paradigm-shifting advances in cancer management. However, their development requires sound preclinical science to address the complexity of clinical trial designs. Certainly, the lessons learned from the recent successful development of other targeted therapeutic agents such as imatinib in CML (Druker 2008) or Hh pathway antagonists in metastatic basal cell carcinoma (Von Hoff et al. 2009) will improve our ability to overcome these challenges.

Acknowledgment This work was supported by NIH5-RO1GM076398 (L.L.)

References

Arber N, Eagle CJ, Spicak J et al (2006) Celecoxib for the prevention of colorectal adenomatous polyps. N Engl J Med 355:885–895

Arce L, Yokoyama NN, Waterman ML (2006) Diversity of LEF/TCF action in development and disease. Oncogene 25:7492–7504

Badiglian Filho L, Oshima CT, De Oliveira Lima F et al (2009) Canonical and noncanonical Wnt pathway: a comparison among normal ovary, benign ovarian tumor and ovarian cancer. Oncol Rep 21:313–320

Baeg GH, Perrimon N (2000) Functional binding of secreted molecules to heparan sulfate proteoglycans in Drosophila. Curr Opin Cell Biol 12:575–580

Bailey JM, Singh PK, Hollingsworth MA (2007) Cancer metastasis facilitated by developmental pathways: Sonic hedgehog, Notch, and bone morphogenic proteins. J Cell Biochem 102:829–839

Banziger C, Soldini D, Schutt C et al (2006) Wntless, a conserved membrane protein dedicated to the secretion of Wnt proteins from signaling cells. Cell 125:509–522

Barker N, Ridgway RA, van Es JH et al (2009) Crypt stem cells as the cells-of-origin of intestinal cancer. Nature 457:608–611

Bartscherer K, Pelte N, Ingelfinger D, Boutros M (2006) Secretion of Wnt ligands requires Evi, a conserved transmembrane protein. Cell 125:523–533

Blanc E, Roux GL, Benard J, Raguenez G (2005) Low expression of Wnt-5a gene is associated with high-risk neuroblastoma. Oncogene 24:1277–1283

Boman BM, Wicha MS (2008) Cancer stem cells: a step toward the cure. J Clin Oncol 26:2795–2799

Brantjes H, Roose J, van De Wetering M, Clevers H (2001) All Tcf HMG box transcription factors interact with Groucho-related co-repressors. Nucleic Acids Res 29:1410–1419

Buchdunger E, Cioffi CL, Law N et al (2000) Abl Protein-Tyrosine Kinase Inhibitor STI571 Inhibits In Vitro Signal Transduction Mediated by c-Kit and Platelet-Derived Growth Factor Receptors. J Pharmacol Exp Ther 295:139–145

Cadigan KM (2008) Wnt-beta-catenin signaling. Curr Biol 18:R943–947

Chen B, Dodge ME, Tang W et al (2009) Small molecule-mediated disruption of Wnt-dependent signaling in tissue regeneration and cancer. Nat Chem Biol 5:100–107

Chien AJ, Conrad WH, Moon RT (2009) A Wnt survival guide: from flies to human disease. J Invest Dermatol 129:1614–1627

Cong F, Zhang J, Pao W et al (2003) A protein knockdown strategy to study the function of beta-catenin in tumorigenesis. BMC Mol Biol 4:10

Couso JP, Martinez Arias A (1994) Notch is required for wingless signaling in the epidermis of Drosophila. Cell 79:259–272

Da Forno PD, Pringle JH, Hutchinson P et al (2008) WNT5A expression increases during melanoma progression and correlates with outcome. Clin Cancer Res 14:5825–5832

de Lau W, Barker N, Clevers H (2007) WNT signaling in the normal intestine and colorectal cancer. Front Biosci 12:471–491

Dejmek J, Dejmek A, Safholm A et al (2005) Wnt-5a protein expression in primary dukes B colon cancers identifies a subgroup of patients with good prognosis. Cancer Res 65:9142–9146

Derksen PW, Tjin E, Meijer HP et al (2004) Illegitimate WNT signaling promotes proliferation of multiple myeloma cells. Proc Natl Acad Sci U S A 101:6122–6127

Druker BJ (2008) Translation of the Philadelphia chromosome into therapy for CML. Blood 112:4808–4817

Emami KH, Nguyen C, Ma H et al (2004) A small molecule inhibitor of beta-catenin/CREB-binding protein transcription [corrected]. Proc Natl Acad Sci U S A 101:12682–12687

Fevr T, Robine S, Louvard D, Huelsken J (2007) Wnt/beta-catenin is essential for intestinal homeostasis and maintenance of intestinal stem cells. Mol Cell Biol 27:7551–7559

Finch PW, He X, Kelley MJ et al (1997) Purification and molecular cloning of a secreted, Frizzled-related antagonist of Wnt action. Proc Natl Acad Sci U S A 94:6770–6775

Giles RH, van Es JH, Clevers H (2003) Caught up in a Wnt storm: Wnt signaling in cancer. Biochim Biophys Acta 1653:1–24

Glinka A, Wu W, Delius H et al (1998) Dickkopf-1 is a member of a new family of secreted proteins and functions in head induction. Nature 391:357–362

Groen RW, Oud ME, Schilder-Tol EJ et al (2008) Illegitimate WNT pathway activation by beta-catenin mutation or autocrine stimulation in T-cell malignancies. Cancer Res 68:6969–6977

Hecht A, Vleminckx K, Stemmler MP et al (2000) The p300/CBP acetyltransferases function as transcriptional coactivators of beta-catenin in vertebrates. Embo J 19:1839–1850

Horvath LG, Henshall SM, Kench JG et al (2004) Membranous expression of secreted frizzled-related protein 4 predicts for good prognosis in localized prostate cancer and inhibits PC3 cellular proliferation in vitro. Clin Cancer Res 10:615–625

Hsieh JC, Kodjabachian L, Rebbert ML et al (1999) A new secreted protein that binds to Wnt proteins and inhibits their activities. Nature 398:431–436

Huang SM, Mishina YM, Liu S et al (2009) Tankyrase inhibition stabilizes axin and antagonizes Wnt signalling. Nature 461:614–620

Jamieson CH, Ailles LE, Dylla SJ et al (2004) Granulocyte-macrophage progenitors as candidate leukemic stem cells in blast-crisis CML. N Engl J Med 351:657–667

Jiang Y, Prunier C, Howe PH (2008) The inhibitory effects of Disabled-2 (Dab2) on Wnt signaling are mediated through Axin. Oncogene 27:1865–1875

Jonsson M, Dejmek J, Bendahl PO, Andersson T (2002) Loss of Wnt-5a protein is associated with early relapse in invasive ductal breast carcinomas. Cancer Res 62:409–416

Kadowaki T, Wilder E, Klingensmith J et al (1996) The segment polarity gene porcupine encodes a putative multitransmembrane protein involved in Wingless processing. Genes Dev 10:3116–3128

Koesters R, Ridder R, Kopp-Schneider A et al (1999) Mutational activation of the beta-catenin proto-oncogene is a common event in the development of Wilms' tumors. Cancer Res 59:3880–3882

Kurayoshi M, Oue N, Yamamoto H et al (2006) Expression of Wnt-5a is correlated with aggressive-ness of gastric cancer by stimulating cell migration and invasion. Cancer Res 66:10439–10448

Liu J, Stevens J, Matsunami N, White RL (2004) Targeted degradation of beta-catenin by chimeric F-box fusion proteins. Biochem Biophys Res Commun 313:1023–1029

Logan CY, Nusse R (2004) The Wnt signaling pathway in development and disease. Annu Rev Cell Dev Biol 20:781–810

Lu W, Yamamoto V, Ortega B, Baltimore D (2004) Mammalian Ryk is a Wnt coreceptor required for stimulation of neurite outgrowth. Cell 119:97–108

Ma H, Nguyen C, Lee KS, Kahn M (2005) Differential roles for the coactivators CBP and p300 on TCF/beta-catenin-mediated survivin gene expression. Oncogene 24:3619–3631

Maier TJ, Janssen A, Schmidt R et al (2005) Targeting the beta-catenin/APC pathway: a novel mechanism to explain the cyclooxygenase-2-independent anticarcinogenic effects of celecoxib in human colon carcinoma cells. FASEB J 19:1353–1355

Malanchi I, Peinado H, Kassen D et al (2008) Cutaneous cancer stem cell maintenance is dependent on beta-catenin signalling. Nature 452:650–653

McDonald SL, Silver A (2009) The opposing roles of Wnt-5a in cancer. Br J Cancer 101:209–214

McMillan M, Kahn M (2005) Investigating Wnt signaling: a chemogenomic safari. Drug Discov Today 10:1467–1474

McWhirter JR, Neuteboom ST, Wancewicz EV et al (1999) Oncogenic homeodomain transcrip-tion factor E2A-Pbx1 activates a novel WNT gene in pre-B acute lymphoblastoid leukemia. Proc Natl Acad Sci U S A 96:11464–11469

Miyagishi M, Fujii R, Hatta M et al (2000) Regulation of Lef-mediated transcription and p53-dependent pathway by associating beta-catenin with CBP/p300. J Biol Chem 275:35170–35175

Nagao R, K.S., Ashihara E, Takeuchi M, et al. . (2008). A novel b-catenin inhibitor, AV65 suppresses the growth of CML cell lines which acquire imatinib-resistance because of Abl kinase domain mutations including T315I and hypoxia-adaptation. Blood. 112:A395

Nguyen DX, Chiang AC, Zhang XH et al (2009) WNT/TCF signaling through LEF1 and HOXB9 mediates lung adenocarcinoma metastasis. Cell 138:51–62

Nusse R, Varmus HE (1982) Many tumors induced by the mouse mammary tumor virus contain a provirus integrated in the same region of the host genome. Cell 31:99–109

Nusse R, Brown A, Papkoff J et al (1991) A new nomenclature for int-1 and related genes: the Wnt gene family. Cell 64:231

Nusse R (2003) Wnts and Hedgehogs: lipid-modified proteins and similarities in signaling mechanisms at the cell surface. Development 130:5297–5305

Nusse R (2005) Wnt signaling in disease and in development. Cell Res 15:28–32

Oishi I, Suzuki H, Onishi N et al (2003) The receptor tyrosine kinase Ror2 is involved in noncanonical Wnt5a/JNK signalling pathway. Genes Cells 8:645–654

Polakis P (2000) Wnt signaling and cancer. Genes Dev 14:1837–1851

Polyak K, Weinberg RA (2009) Transitions between epithelial and mesenchymal states: acquisition of malignant and stem cell traits. Nat Rev Cancer 9:265–273

Rao AS, Kremenevskaja N, von Wasielewski R et al (2006) Wnt/beta-catenin signaling mediates antineoplastic effects of imatinib mesylate (gleevec) in anaplastic thyroid cancer. J Clin Endocrinol Metab 91:159–168

Reed KR, Athineos D, Meniel VS et al (2008) B-catenin deficiency, but not Myc deletion, suppresses the immediate phenotypes of APC loss in the liver. Proc Natl Acad Sci U S A 105:18919–18923

Reya T, Clevers H (2005) Wnt signalling in stem cells and cancer. Nature 434:843–850

Rijsewijk F, Schuermann M, Wagenaar E et al (1987) The Drosophila homolog of the mouse mammary oncogene int-1 is identical to the segment polarity gene wingless. Cell 50:649–657

Segditsas S, Sieber OM, Rowan A et al (2008) Promoter hypermethylation leads to decreased APC mRNA expression in familial polyposis and sporadic colorectal tumours, but does not substitute for truncating mutations. Exp Mol Pathol 85:201–206

Sharma, R. (1973). Wingless, a new mutant in D. melanogaster *Dros Inf Serv*. 50:134

Su Y, Ishikawa S, Kojima M, Liu B (2003) Eradication of pathogenic beta-catenin by Skp1/Cullin/F box ubiquitination machinery. Proc Natl Acad Sci U S A 100:12729–12734

Sun Y, Kolligs FT, Hottiger MO et al (2000) Regulation of beta -catenin transformation by the p300 transcriptional coactivator. Proc Natl Acad Sci U S A 97:12613–12618

Takemaru KI, Moon RT (2000) The transcriptional coactivator CBP interacts with beta-catenin to activate gene expression. J Cell Biol 149:249–254

Tang W, Dodge M, Gundapaneni D et al (2008) A genome-wide RNAi screen for Wnt/beta-catenin pathway components identifies unexpected roles for TCF transcription factors in cancer. Proc Natl Acad Sci U S A 105:9697–9702

Tejpar S, Nollet F, Li C et al (1999) Predominance of beta-catenin mutations and beta-catenin dysregulation in sporadic aggressive fibromatosis (desmoid tumor). Oncogene 18:6615–6620

Urakami S, Shiina H, Enokida H et al (2006) Epigenetic inactivation of Wnt inhibitory factor-1 plays an important role in bladder cancer through aberrant canonical Wnt/beta-catenin signaling pathway. Clin Cancer Res 12:383–391

van Amerongen R, Mikels A, Nusse R (2008) Alternative wnt signaling is initiated by distinct receptors. Sci Signal 1:1–5

van den Heuvel M, Harryman-Samos C, Klingensmith J et al (1993) Mutations in the segment polarity genes wingless and porcupine impair secretion of the wingless protein. Embo J 12: 5293–5302

van Stolk R, Stoner G, Hayton WL et al (2000) Phase I trial of exisulind (sulindac sulfone, FGN-1) as a chemopreventive agent in patients with familial adenomatous polyposis. Clin Cancer Res 6:78–89

Vincan E, Barker N (2008) The upstream components of the Wnt signalling pathway in the dynamic EMT and MET associated with colorectal cancer progression. Clin Exp Metastasis 25:657–663

Von Hoff DD, LoRusso PM, Rudin CM et al (2009) Inhibition of the hedgehog pathway in advanced basal-cell carcinoma. N Engl J Med 361:1164–1172

Willert K, Brown JD, Danenberg E et al (2003) Wnt proteins are lipid-modified and can act as stem cell growth factors. Nature 423:448–452

Yoshizumi T, Ohta T, Ninomiya I et al (2004) Thiazolidinedione, a peroxisome proliferator-activated receptor-gamma ligand, inhibits growth and metastasis of HT-29 human colon cancer cells through differentiation-promoting effects. Int J Oncol 25:631–639

Chapter 7
STAT Signaling in the Pathogenesis and Treatment of Cancer

Sarah R. Walker and David A. Frank

7.1 Introduction

Since the beginning of the chemotherapy era in the 1940s, the approach to anti-cancer drug development has centered around the identification of compounds that damaged key components of cellular function. From alkylating agents, to antimetabolites, microtubule poisons, and topoisomerase inhibitors, the therapeutic approach was predicated on inducing cytotoxicity with the expectation that rapidly dividing cancer cells would be affected to a greater extent than normal cells. As inelegant as this approach may now seem today, it did lead to the ability to cure a high percentage of patients with childhood acute lymphoblastic leukemia (ALL), Hodgkin lymphoma, and testicular cancer, and enhanced the outcome in many other forms of cancer. However, this approach carried with it a large burden of side effects, including the increased risk of second malignancies.

During the last 40 years, our molecular understanding of the events driving the malignant behavior of the cancer cell has increased enormously. While it seems that the benefit that can be derived from cytotoxic therapy has reached a plateau, the opportunity to translate our molecular understanding into rational therapies is greater than ever. Much of the focus on rational molecular therapy has been on sequencing tumor DNA to identify mutations that are potential "drivers" of the malignant phenotype of the cancer cell. Based on findings in diseases such as chronic myelogenous leukemia (CML), much of the initial focus has been on identifying mutations in kinases, particularly tyrosine kinases, whether they arise from chromosomal translocations, point mutations, deletions, or internal tandem duplications. However, it is becoming clear that such activating mutations in kinases are not as common as

S.R. Walker • D.A. Frank (✉)
Department of Medical Oncology, Dana Farber Cancer Institute, Harvard Medical School, 450 Brookline Avenue, Boston, MA 02215, USA
e-mail: david_frank@dfci.harvard.edu

D.A. Frank (ed.), *Signaling Pathways in Cancer Pathogenesis and Therapy*,
DOI 10.1007/978-1-4614-1216-8_7, © Springer Science+Business Media, LLC 2012

anticipated in many cancers. Furthermore, with the exception of notable successes such as CML, even where such mutations have been found in kinases, the response to kinase inhibitors has been incomplete or of only transient benefit.

These findings raise the possibility of developing an alternate approach to personalized molecular therapy of tumors. Although there are hundreds of kinases that can become activated in tumorigenesis, all of these pathways must converge on a relatively small set of genes that regulate key cellular functions such as cell cycle progression, survival, and self-renewal. Thus, one can work upstream from these genes to identify key transcription factors whose activation is central to maintaining the malignant state. While these transcription factors themselves may not be mutated, their dysregulation leads to the persistent expression of genes necessary for malignancy. Understanding these oncogenic transcription factors can therefore provide insight into the pathogenesis of these tumors, and also suggest novel therapeutic strategies as well.

7.2 STAT Transcription Factors

Cancer cells subvert normal physiologic signaling pathways to drive their malignant phenotype, which at its core requires proliferation, survival, and self-renewal. Under physiological conditions, these phenotypes are driven by extracellular cues, principally growth factors and cytokines. In the early 1990s, a family of seven homologous transcription factors that could transduce signals from these stimuli was identified. These proteins, now called STATs, resided in the cytoplasm under basal conditions. Once phosphorylated on a single tyrosine residue near the carboxyl terminus, STAT dimers translocate to the nucleus where they bind to specific DNA sequences and activate transcription (Fig. 7.1). As with many transcriptional modulators, closer study has revealed that these proteins can repress the expression of certain target genes as well, and understanding how these proteins manage these seemingly opposite effects is an area of great interest (Walker et al. 2007). Reflecting the fact that the target genes regulated by STATs control critical cellular processes, STAT activation is very transient. STATs are dephosphorylated and exported from the nucleus or degraded within minutes to a few hours, allowing a tight control of gene regulation to be maintained. STAT activity is controlled by phosphatases, SOCS proteins that prevent subsequent STAT phosphorylation, and PIAS family members that lead to the inactivation of phosphorylated STATs often through sumoylation (Jackson 2001). In fact, many STAT target genes are themselves negative regulators of the STAT signaling pathway.

While the basic transcriptional function of STATs has focused on STAT tyrosine phosphorylation as the key regulatory event, there are clearly additional subtleties to STAT signaling that are still being elucidated. For example, STATs can also be phosphorylated on specific serine residues as well, by a range of serine/threonine kinases. Serine phosphorylation may modulate the tyrosine phosphorylation and transcriptional function of STATs (Chung et al. 1997; Jain et al. 1998). However, it

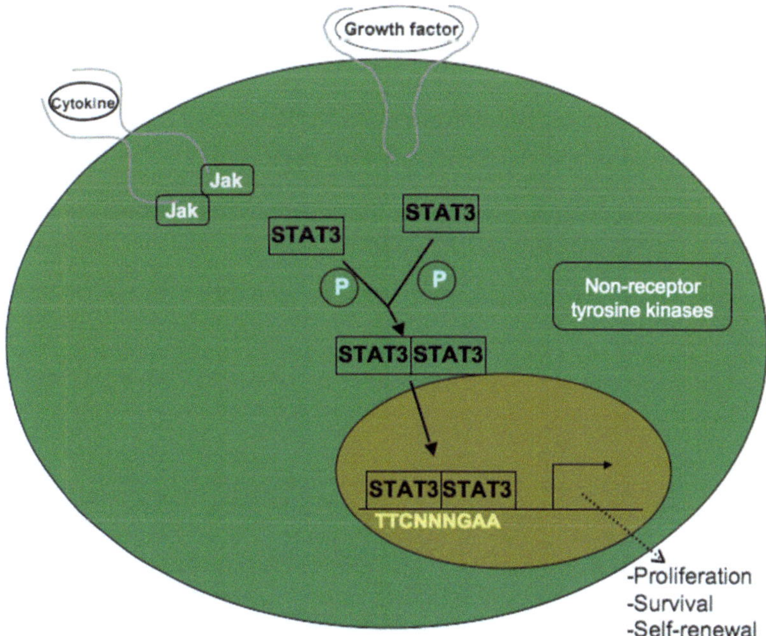

Fig. 7.1 STAT signal transduction. STATs, exemplified here by STAT3, are latent transcription factors found predominantly in the cytoplasm under basal conditions. STATs are activated by phosphorylation on a single tyrosine residue, which can be triggered by Jak kinases associated with cytokine receptors, growth factor receptor tyrosine kinases, and non-receptor tyrosine kinases. Phosphorylated STAT dimers translocate to the nucleus, bind to specific nine base pair DNA sequences in the regulatory regions of target genes, and activate (or repress) transcription. STAT3 target genes regulate cellular processes such as proliferation, survival, and self-renewal. Constitutive activation of STAT3, by driving continuous high level expression of these target genes, can mediate oncogenic effects

is becoming increasingly clear that serine phosphorylation alone may be sufficient under some circumstances to mediate transcriptional effects independent of tyrosine phosphorylation (Hazan-Halevy et al. 2010). Furthermore, STAT3 serine phosphorylation has been shown to have significant effects in regulating mitochondrial function (Gough et al. 2009). Furthermore, STATs can be regulated by acetylation and other posttranslational modifications, and thus the full spectrum of STAT-mediated biological effects is still being elucidated. Unphosphorylated STAT3 may also associate with other transcription factors to modulate transcription, and may alter cellular function through nontranscriptional mechanism including interactions with the cytoskeleton (Germain and Frank 2007; Walker et al. 2010; Yang and Stark 2008).

However, the finding that STATs coordinately regulate genes that control the processes central to tumorigenesis raised the possibility that aberrant activation of these proteins could be a critical event in the molecular pathogenesis of human cancers. The mammalian STAT family contains seven members, including STAT5a and STAT5b, which are two highly related but distinct STATs that function largely interchangeably. Although there may be subtle differences in their expression or their function, they are often grouped together as STAT5. STAT5 and STAT3 are widely expressed, are activated downstream of a variety of stimuli, and generally mediate effects of stimuli promoting cellular proliferation. This is in distinction to the other STAT family members which play more restricted roles. STAT1 and STAT2 mediate the effects of interferons, and generally mediate an antiproliferative effect. STAT4 and STAT6 are largely involved in the differentiation and functioning of T lymphocytes. While the other STAT family members may play a role in certain malignancies, the focus from an oncology standpoint has largely been on STAT3 and STAT5.

7.3   STAT Activation in Cancer

Target genes of STAT3 and STAT5 regulate processes such as cell cycle progression (e.g., cyclin D1), survival (e.g., Mcl-1, Bcl-xl, and Bcl-2), and self-renewal (e.g., Bcl-6 and Klf-4). While physiologic STAT activation is rapid and transient, it is clear that constitutive activation of these transcription factors leads to the high level expression of genes that can generate a neoplastic phenotype. Immunohistochemistry (IHC) can be performed on tissue to detect the tyrosine phosphorylated, activated forms of these STATs. In normal tissue, an occasional nucleus may be detected with an activated form of STAT3 or STAT5. By contrast, in a wide range of human malignancies, constitutive phosphorylation of one of these STAT family members can be found in the majority of malignant cells. For example, activation of STAT3, as detected by IHC, immunoblots, or electrophoretic mobility shift assays (EMSAs) has been found to occur commonly in a range of cancers including melanoma and cancers of the breast, colon, lung, ovary, pancreas, prostate, esophagus, and stomach. In breast cancer, activation of STAT5 (which mediates the effects of prolactin) has also been described. This can occur in conjunction with activation of STAT3 or independent of STAT3, and the molecular and clinical characteristics of breast cancers differ depending on the relative activation of these two proteins (Walker et al. 2009).

Hematological cancers also commonly display activation of STAT3 and/or STAT5. STAT5 activation is nearly universal in CML, and STAT3 activation is more common in multiple myeloma. In acute myeloid leukemia (AML), both STAT5 and STAT3 can be activated. The mechanisms restricting the activation of these various STAT family members, and the effects on gene expression and biology of these tumors remain an important area of research.

Several additional variations on STAT activation have been found. For example, in chronic lymphocytic leukemia (CLL), the most common leukemia in developed countries, STAT3, is universally phosphorylated on the carboxyl terminal serine residue, though not on the regulatory tyrosine residue (Frank et al. 1997). Nonetheless, STAT3-dependent gene expression occurs in these cells, and they appear dependent on STAT3 for survival (Hazan-Halevy et al. 2010). Although STAT6 activation is not generally associated with malignancy, several forms of lymphoma have been reported to be characterized by constitutive phosphorylation of this family member (Bruns and Kaplan 2006; Guiter et al. 2004).

These findings raise several key questions. The first issue is whether the phosphorylation of these STATs is "driving" the pathogenesis of these cancers, or is merely a "passenger" reflecting the activation of some upstream kinase pathway, but not directly affecting the biology of the cancer. Abundant evidence derived from cell culture experiments and genetic models supports the pathogenic role of STATs in the pathogenesis of these tumors. Specifically, inhibition of constitutively activated STATs, by genetic deletion, RNA interference, antisense inhibition, or pharmacological antagonists, leads to a loss of STAT-dependent gene expression, and decreased tumor cell survival (Frank 2006; Frank 2007). Conversely, artificial mutants of STAT3 that are constitutively active can lead to malignant transformation of fibroblasts, suggesting that target genes turned on by STAT3 are sufficient on their own for tumor development (Bromberg et al. 1999) (Alvarez et al. 2005). A constitutively active form of STAT3 has not been reported to be found in human cancers, indicating that STAT3 acts as an oncogenic transcription factor not through its own mutation, but rather by conveying signals generated via mutations at upstream signaling points.

A second question emerging from these findings concerns the mechanism by which STATs become activated in cancer. For some cancers, constitutive STAT phosphorylation is clearly driven by a mutated kinase. One of the earliest and clearest examples of this mechanism was found in CML. The leukemic cells in essentially every patient with CML have a translocation between chromosomes 9 and 22 yielding a chimeric protein, Bcr/Abl. Whereas c-Abl is a nuclear tyrosine kinase that responds to DNA damage, Bcr/Abl is a highly active cytoplasmic kinase that phosphorylates a large number of cellular substrates, including STAT5. Inhibition of Bcr/Abl kinase activity rapidly shuts off STAT5 phosphorylation. Although additional kinases may play a cooperating role, it is clear that Bcr/Abl is driving the phosphorylation of STAT5, and that STAT5-depedent gene expression is critical for the pathogenesis of this disease. Similarly, a number of other mutated kinases can drive the inappropriate phosphorylation of STATs.

However, the frequency of STAT activation in many tumors exceeds the frequency of mutated kinases identified in those cancers. For example, approximately 70% of prostate cancers display constitutive activation of STAT3 (Alvarez et al. 2005). However, genomic sequencing of prostate cancers has found relatively few examples of tyrosine kinases that have been activated by mutations. In breast cancer, approximately 75% of tumors display activation of STAT3 and/or STAT5 (Walker et al. 2009). The most commonly mutated kinase, Her2/ErbB2, is activated by amplification in approximately 25–30% of breast cancers, and relatively few other activated kinases have been identified in breast cancer. These findings have led to two significant observations. First, negative regulators of STAT activation, including phosphatases and SOCS family members, are commonly inactivated in cancers displaying constitutive STAT activation, often through promoter methylation (Chim et al. 2004; Galm et al. 2003). This should not be surprising in that many of the target genes regulated by STATs are negative regulators of this pathway that serve to attenuate signaling. However, in many systems, loss of negative regulators alone can lead to constitutive STAT activation, suggesting that STAT phosphorylation and dephosphorylation likely occurs under basal conditions, and loss of the negative regulatory arm can "trap" STATs in the phosphorylated state.

A second key finding emerged when it was found that a common mechanism driving the phosphorylation of STATs in tumors ranging from multiple myeloma to breast cancer is autocrine production of cytokines, particularly interleukin (IL)-6 (Kawano et al. 1988; Marotta et al. 2011; Szczepek et al. 2001). This observation leads to several important implications. First, it reinforces the idea that an activating mutation in a kinase need not be present to drive the activation of STATs. It also suggests that non-mutated kinases, such as the Jaks that transduce signals from cytokine receptors, might be important therapeutic targets under these conditions. In addition, since cytokine signaling normally leads to only a transient response of STAT phosphorylation, it suggests that defects in negative feedback regulators likely are of importance even in the presence of these autocrine loops. Finally, the finding that local cytokine production can activate STAT3 in tumor cells raises another important observation. Nonmalignant cells in the vicinity of the tumor, particularly immune cells, can also have STAT3 activated by these cytokines released from the tumor cells, or in some cases from stromal cells (Uchiyama et al. 1993). This is of considerable importance given that the activation of STAT3 in antigen processing cells and other immune cells leads to an immunosuppressive effect (Wang et al. 2004). Thus, STAT3 activation can both drive the malignant behavior of cancer cells as well as suppress the immune response to the cancer. However, it also suggests that pharmacological inhibitors of STAT3 can provide benefit both through their direct anticancer effects and immune stimulatory actions.

As noted, STAT target genes can promote cell cycle progression, survival (including resistance to apoptosis induced by cytotoxic drugs and radiation therapy) and self-renewal (or the maintenance of cancer stem cells). In addition, STAT target genes such as matrix metalloproteinases (MMPs) and vascular endothelial growth factor (VEGF) can promote invasion and angiogenesis, respectively. Thus, constitutive activation of STATs can promote both cell autonomous phenotypes associated with malignancy as well as phenotypes associated with metastasis.

7.4 STAT Inhibition as a Therapeutic Strategy in Cancer

Taken together, there is strong evidence that STATs, particularly STAT3 and STAT5, are commonly activated in a range of human cancers, and directly drive the malignant phenotype of these cells. In considering STATs as a target for therapy, the first critical issue concerns the question of what will happen to normal cells if a STAT family member is inhibited. Data from a range of systems suggests that loss of STAT3 or STAT5 in adult tissue can be tolerated without deleterious effects. In fact, familial forms of the hyper-IgE syndrome have been mapped to STAT3, and the mutations described are predicted to function as dominant inhibitory forms of STAT3 (Holland et al. 2007). Since patients with this syndrome develop normally and are fertile, this "experiment of nature" suggests that inhibiting STAT3 in an adult should be tolerated well by normal cells. These findings lend credence to the concept that specific STAT inhibitors can have a high therapeutic index in treating cancers.

The second issue that emerges in considering STATs as therapeutic targets is whether it is possible to develop pharmacological strategies to inhibit these proteins. Transcription factors have traditionally been viewed as difficult targets for designing inhibitors. The surfaces of their protein–protein and protein–DNA interactions are relatively large and flat, making it difficult to design small molecules that can inhibit them effectively. However, despite these concerns, great progress has been made in identifying ways to target STATs, particularly STAT3. Some drugs identified as STAT inhibitors function by inhibiting upstream kinases that are driving the tyrosine phosphorylation of STATs, as well as other cellular substrates. Others, particularly a range of natural products, seem to inhibit a variety of pathways simultaneously, often at high concentrations. However, using rational design or cell-based screening systems, a number of more specific STAT inhibitors have emerged, and clinical trials are beginning with some of these agents, as described below.

7.5   Approaches to the Development of STAT Inhibitors

7.5.1   Rational Design

7.5.1.1   SH2 Inhibitors

The increasing structural and biological understanding of STAT signaling has afforded two principal opportunities for targeting specific components of STATs, specifically the SH2 domain and the DNA-binding domain. Given the prominent and widespread role of STAT3 in human cancers, most of the pharmacological development has focused on this STAT family member.

Targeting the SH2 domain, which allows proteins to bind to other proteins containing a phosphorylated tyrosine residue, is particularly appealing, since it is necessary for two key steps in STAT activation, recruitment to an activated receptor–kinase complex (which itself becomes tyrosine phosphorylated) and dimerization (Turkson et al. 2004; Turkson et al. 2001). In tumor cells, STAT3 signals largely as homodimers, though STAT3 may also form heterodimers with STAT1. Much of the initial work on developing inhibitors of the STAT3 SH2 domain started with a phosphopeptide backbone that mimicked the sequence around the STAT3 phosphorylated tyrosine residue (tyrosine 705) which is its physiologic binding partner. Since phosphopeptides themselves are poor drug candidates, owing to their poor stability and low cellular permeability, efforts were then made to make nonpeptide analogs with more "drug-like" properties, and continual improvements in the potency of these compounds have been made (Turkson et al. 2004).

An alternate approach to blocking the STAT3 SH2 domain has been to rely on computational approaches to screen *in silico* hundreds of thousands of compounds for their predicted ability to bind to this site with high affinity and specificity. This approach led to the identification of STA-21, or deoxytetrangomycin (Song et al. 2005). This compound was then shown to inhibit STAT3 dimerization, and kill breast cancer cells characterized by constitutive STAT3 activation. STA-21 or its structural analog LLL-3 subsequently has been shown to promote cell killing in bladder cancer, sarcoma, and glioblastoma models, suggesting that this compound family may be functional in multiple cancer types (Chen et al. 2008; Chen et al. 2007; Fuh et al. 2009).

A number of unique compounds have been found to interfere with STAT3 SH2 function. For example, novel guanine-rich oligonucleotides that form structures called G-quartets, which were originally designed to block other steps in signaling, appear to function by blocking the SH2 domain (Jing et al. 2004). Although these compounds have notable activity in vitro, the ultimate clinical applicability of these compounds is unclear.

7.5.1.2 DNA Binding Inhibitors

Since the ultimate transcriptional function of STAT3 is mediated by its binding to a specific nine base pair DNA sequence in the regulatory region of target genes, a strategy for developing STAT3 inhibitors is to disrupt this interaction. One of the most direct ways to achieve this is to use short double stranded oligonucleotides containing a STAT binding site, the so-called decoy oligonucleotides, which can serve as an intracellular sink for activated STATs, thereby attenuating their transcriptional function (Chan et al. 2004; Wang et al. 2000; Xi et al. 2005). In both cell culture systems and xenografts in mice, this approach has been shown to block STAT3-dependent gene expression, and has a therapeutic effect (Boehm et al. 2008; Lui et al. 2007; Sen et al. 2009). As with all nucleic acid-based therapeutics, two key hurdles are maintaining stability and crossing cell membranes which may limit their clinical applicability. However, squamous cell carcinomas of the head and neck commonly manifest constitutive STAT3 activation, and they are generally amenable to direct visualization and intervention. Thus, they provide an ideal tumor type to evaluate intratumoral injections of decoy oligonucleotides. From studies done in primate models, this approach appears feasible, and studies in humans are being initiated (Sen et al. 2009).

Several other approaches to inhibiting STAT3 DNA binding are being developed. For example, screening approaches have been used to identify peptide aptamers that block STAT3 DNA binding (Leeman et al. 2006). However, the challenge will be to synthesize nonpeptide analogs to avoid the issues of stability and cellular permeability that hinder the development of these types of peptide-based therapeutics. STAT3–DNA binding can theoretically be blocked from the DNA surface, using approaches such as polyamides that can bind to DNA in a sequence-specific manner, thereby blocking the function of a transcription factor (Trauger et al. 1996). Finally, recruitment of co-activators by STAT3 might be blocked by specific small molecules, thereby inhibiting the ability of STAT3 to initiate transcription.

7.5.2   Screening Strategies

Given that transcription factors are difficult targets for classical pharmacological intervention, an alternate approach is to develop screening systems to identify compounds that can modulate these proteins. Both cell-based and cell-free screens have been generated in which the activity of STAT3 can be assessed in a high-throughput manner, and libraries of diverse chemicals can be evaluated for their effects. Cell-free systems have the advantage that very specific effects on defined targets can be assessed, and issues related to cellular permeability or nonspecific toxicity can be addressed with further medicinal chemistry efforts. For example, in vitro screens have been used successfully to identify compounds that block STAT3 tyrosine phosphorylation (Fletcher et al. 2009).

While the complexity of cell-based screens is higher, these assays have the advantage that issues related to nonspecific toxicity or cellular permeability can be eliminated at the outset. Furthermore, particularly with the complexity of the steps involved in STAT activation, cell-based assays provide an opportunity to identify multiple potential steps in their activation pathway that can be modulated by small molecules. In one such approach, a luciferase reporter gene is ligated downstream of a STAT-dependent promoter, and then the construct is stably transfected into a cell line (Lynch et al. 2007) (Fig. 7.2). Treatment of these reporter cell lines with cytokines such as IL-6 can then lead to the production of luciferase, which can be detected by luminometry in a high-throughput format. This STAT3 reporter cell line can then be used to evaluate large chemical libraries for molecules that inhibit STAT3 function. To exclude nonspecific effects, a parallel cell line in which luciferase is driven by a related transcription factor, such as STAT5, or an unrelated transcription factor, such as NF-κB, can then be screened in parallel. In this way, only compounds that have a specific effect on STAT3-dependent gene expression will be revealed. A potential challenge of this approach is that although it can identify potent and specific inhibitors, discerning their precise cellular target may be difficult.

One successful use of this strategy has been the identification of the drug nifuroxazide as a STAT3 inhibitor (Nelson et al. 2008). Nifuroxazide, which decreases STAT3 phosphorylation through inhibition of the kinases Jak2 and Tyk2, inhibits STAT3-dependent gene expression, and decreases the viability of multiple myeloma cells dependent on STAT3 activation. The survival of multiple myeloma cells is enhanced by the presence of bone marrow stromal cells, both through the secretion of cytokines as well as direct cell–cell interactions. However, nifuroxazide is able to decrease the viability of myeloma cells even in the presence of these stromal cells. This work thus demonstrates the usefulness of this screening method and lends further proof that STAT3 activation is not a "passenger" but a "driver" in the pathogenesis of cancer.

A high-throughput screening strategy has also been fruitful in identifying the drug pimozide as an inhibitor of STAT5 (Nelson et al. 2011b). Pimozide decreases STAT5 phosphorylation in models of CML, although notably, it is not a kinase inhibitor. While its mechanism of action remains largely unknown, it appears that

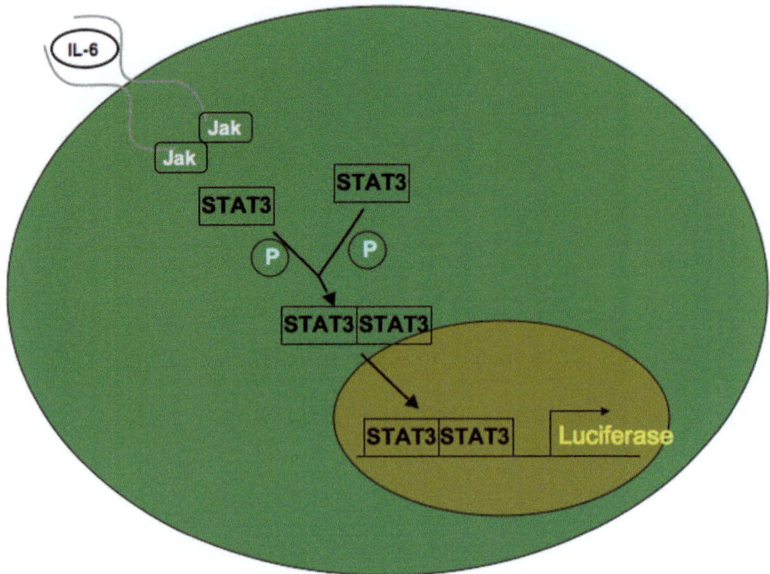

Fig. 7.2 Cell-based screening strategies for STAT3 inhibitors. To identify compounds that inhibit STAT3 by any mechanism, a luciferase reporter construct under the control of a STAT3-responsive element can be stably transfected into a cell line. When cells are treated with IL-6, luciferase is produced in a STAT3-dependent manner, and this can be quantitated by luminometry. Chemical libraries can then be screened for compounds that inhibit STAT3-dependent luciferase production, but which do not affect luciferase production in a parallel cell line in which it is under the control of an unrelated promoter. This strategy can detect inhibitors of STAT phosphorylation, dimerization, nuclear translocation, DNA binding, or co-activator recruitment

pimozide may act through negative regulators of STAT phosphorylation. Nonetheless, pimozide decreases the expression of STAT5 target genes, and decreases the survival of both CML cell lines and primary cells from patients with CML. Although CML can be treated effectively with kinase inhibitors that block the function of Bcr/Abl, resistance can emerge due to the development of point mutations in Bcr/Abl rendering it resistant to pharmacological inhibitors. As would be expected of a drug that targets a downstream mediator, such as STAT5, pimozide shows equal efficacy even in the presence of these mutant forms of Bcr/Abl. Another important finding to emerge from this study concerns the issue of therapeutic index. Many models of leukemia have shown a requirement for STAT5 for leukemogenesis (Levy and Gilliland 2000; Schwaller et al. 2000). STAT5 also plays a role in normal hematopoiesis, although due to redundancies in signaling downstream of cytokines, it is not absolutely essential (Liu et al. 1997; Socolovsky et al. 2001). Thus, a question arises as to whether a STAT5 inhibitor might have significant hematological toxicity. However, while pimozide completely inhibits the ability of CD34+ hematopoietic stem cells from CML patients to generate colonies in vitro, it has little effect on the ability of CD34+ cells from healthy individuals. These findings lend further support to the concept that STAT5 inhibitors in particular, and STAT inhibitors in general, may have limited toxicity.

Cell-based screens have proven to be very effective at identifying compounds that are specific STAT inhibitors, and have revealed steps in STAT activation that may be suitable for further drug development. Since the ultimate goal of these studies is to identify STAT inhibitors that can be introduced into the care of cancer patients, they can also be useful in accelerating proof-of-concept clinical trials. Specifically, one can screen libraries of compounds that have been approved for use in humans for other indications or which are already known to be safe in humans. This can reveal hitherto unknown STAT inhibitory function of these compounds, and given their known safety profile, the hurdles to initiate a clinical trial are significantly reduced. Through this approach, the antiparasitic compound pyrimethamine was found to have STAT3 inhibitory effects (Nelson et al. 2011a). Since STAT3 plays an important role in the pathogenesis of CLL (Frank et al. 1997; Hazan-Halevy et al. 2010), a clinical trial has been initiated in treating CLL patients whose disease has progressed on standard therapies, with pyrimethamine (ClinicalTrials.gov Identifier NCT01066663). Since the malignant cell in CLL is obtainable by venipuncture, this protocol will also allow monitoring of the concentration of pyrimethamine in the CLL cells as well as the effect of pyrimethamine on STAT3-dependent gene expression in these cells. These types of pharmacokinetic and pharmacodynamic analyzes are increasingly important in studies of signal transduction inhibitors, as they allow investigators to answer the key question of whether a drug is actually inhibiting the intended target. If the target, STAT3 in this case, is not inhibited, it will allow an assessment of whether the issue is insufficient blood levels of the drug or another pharmacokinetic factor. If the target is inhibited, but a therapeutic effect is not observed, it raises the question of whether it might be more effective to combine a STAT3 inhibitor with another therapeutic agent. It is clear that to move these targeted approaches from the laboratory to the clinic, these types of scientific studies must be intrinsic to the clinical trial.

Finally, it should be recognized that single agent STAT inhibitors may not be optimal anti-cancer agents. Although STAT3 target genes recapitulate the spectrum of phenotypes driving a malignant cell, a STAT3 inhibitor may not be sufficient on its own since inhibition of one signaling pathway may lead to compensatory effects in other signaling pathways. For example, the STAT5 inhibitor pimozide leads to activation of the Erk MAP kinases. However, this can be exploited therapeutically by combining a STAT5 inhibitor and a MEK inhibitor (Nelson et al. 2011b). In addition, although STAT inhibitors may decrease expression of pro-survival genes, this may not be sufficient to induce apoptosis, but may merely lower the threshold for apoptosis. Therefore, analyzing the effects of STAT inhibitors by techniques such as BH3 profiling, which measures apoptotic priming, may allow a deeper understanding of how these agents can be optimized clinically. For example, it may be that a STAT inhibitor may reduce resistance to cytotoxic agents or ionizing radiation, and may best be used in conjunction with these standard therapies.

7.6 Conclusion

As the ability to sequence the genome of every patient's cancer cells comes closer to reality, the hope is that we will be able to choose a specific targeted therapy to neutralize the pathogenic mutations. For activating mutations in kinases, we are coming closer to that ability. However, it is becoming increasingly clear that given the complexity of the cancer genome, it may be a long time before that approach is feasible. In the shorter term, it may be useful to identify the activation of transcription factors, such as STATs, that sit at a convergence point of multiple pathways and drive the gene expression underlying malignant cellular behavior. This type of phenotypic characterization can then be exploited with a range of novel targeting approaches that have shown the potential of providing a high therapeutic index in a wide range of human cancers.

References

Alvarez JV, Febbo PG, Ramaswamy S, Loda M, Richardson A, Frank DA (2005) Identification of a genetic signature of activated signal transducer and activator of transcription 3 in human tumors. Cancer Res 65:5054–5062

Boehm AL, Sen M, Seethala R, Gooding WE, Freilino M, Wong SM, Wang S, Johnson DE, Grandis JR (2008) Combined targeting of epidermal growth factor receptor, signal transducer and activator of transcription-3, and Bcl-X(L) enhances antitumor effects in squamous cell carcinoma of the head and neck. Mol Pharmacol 73:1632–1642

Bromberg JF, Wrzeszczynska MH, Devgan G, Zhao Y, Pestell RG, Albanese C, Darnell JE Jr (1999) *Stat3* as an oncogene. Cell 98:295–303

Bruns HA, Kaplan MH (2006) The role of constitutively active Stat6 in leukemia and lymphoma. Crit Rev Oncol Hematol 57:245–253

Chan KS, Sano S, Kiguchi K, Anders J, Komazawa N, Takeda J, DiGiovanni J (2004) Disruption of Stat3 reveals a critical role in both the initiation and the promotion stages of epithelial carcinogenesis. J Clin Invest 114:720–728

Chen CL, Loy A, Cen L, Chan C, Hsieh FC, Cheng G, Wu B, Qualman SJ, Kunisada K, Yamauchi-Takihara K, Lin J (2007) Signal transducer and activator of transcription 3 is involved in cell growth and survival of human rhabdomyosarcoma and osteosarcoma cells. BMC Cancer 7:111

Chen CL, Cen L, Kohout J, Hutzen B, Chan C, Hsieh FC, Loy A, Huang V, Cheng G, Lin J (2008) Signal transducer and activator of transcription 3 activation is associated with bladder cancer cell growth and survival. Mol Cancer 7:78

Chim CS, Fung TK, Cheung WC, Liang R, Kwong YL (2004) SOCS1 and SHP1 hypermethylation in multiple myeloma: implications for epigenetic activation of the Jak/STAT pathway. Blood 103:4630–4635

Chung J, Uchida E, Grammer TC, Blenis J (1997) STAT3 serine phosphorylation by ERK-dependent and -independent pathways negatively modulates its tyrosine phosphorylation. Mol Cell Biol 17:6508–6516

Fletcher S, Drewry JA, Shahani VM, Page BD, Gunning PT (2009) Molecular disruption of oncogenic signal transducer and activator of transcription 3 (STAT3) protein. Biochem Cell Biol 87:825–833

Frank DA (2006) STAT inhibition in the treatment of cancer: transcription factors as targets for molecular therapy. Curr Cancer Therapy Reviews 2:57–65

Frank DA (2007) STAT3 as a central mediator of neoplastic cellular transformation. Cancer Lett 251:199–210

Frank DA, Mahajan S, Ritz J (1997) B lymphocytes from patients with chronic lymphocytic leukemia contain signal transducer and activator of transcription (STAT) 1 and STAT3 constitutively phosphorylated on serine residues. J Clin Invest 100:3140–3148

Fuh B, Sobo M, Cen L, Josiah D, Hutzen B, Cisek K, Bhasin D, Regan N, Lin L, Chan C et al (2009) LLL-3 inhibits STAT3 activity, suppresses glioblastoma cell growth and prolongs survival in a mouse glioblastoma model. Br J Cancer 100:106–112

Galm O, Yoshikawa H, Esteller M, Osieka R, Herman JG (2003) SOCS-1, a negative regulator of cytokine signaling, is frequently silenced by methylation in multiple myeloma. Blood 101:2784–2788

Germain D, Frank DA (2007) Targeting the cytoplasmic and nuclear functions of STAT3 for cancer therapy. Clin Cancer Res 13:5665–5669

Gough DJ, Corlett A, Schlessinger K, Wegrzyn J, Larner AC, Levy DE (2009) Mitochondrial STAT3 supports Ras-dependent oncogenic transformation. Science 324:1713–1716

Guiter C, Dusanter-Fourt I, Copie-Bergman C, Boulland M-L, le Gouvello S, Gaulard P, Leroy K, Castellano F (2004) Constitutive STAT6 activation in primary mediastinal large B-cell lymphoma. Blood 104:543–549

Hazan-Halevy I, Harris D, Liu Z, Liu J, Li P, Chen X, Shanker S, Ferrajoli A, Keating MJ, Estrov Z (2010) STAT3 is constitutively phosphorylated on serine 727 residues, binds DNA, and activates transcription in CLL cells. Blood 115:2852–2863

Holland SM, DeLeo FR, Elloumi HZ, Hsu AP, Uzel G, Brodsky N, Freeman AF, Demidowich A, Davis J, Turner ML et al (2007) STAT3 mutations in the hyper-IgE syndrome. N Engl J Med 357:1608–1619

Jackson PK (2001) A new RING for SUMO: wrestling transcription responses into nuclear bodies with PIAS family E3 SUMO ligases. Genes Dev 15:3053–3058

Jain N, Zhang T, Fong SL, Lim CP, Cao X (1998) Repression of Stat3 activity by activation of mitogen-activated protein kinase (MAPK). Oncogene 17:3157–3167

Jing N, Li Y, Xiong W, Sha W, Jing L, Tweardy DJ (2004) G-quartet oligonucleotides: a new class of signal transducer and activator of transcription 3 inhibitors that suppress growth of prostate and breast tumors through induction of apoptosis. Cancer Res 64:6603–6609

Kawano M, Hirano T, Matsuda T, Taga T, Horii Y, Iwato K, Asaoku H, Tang B, Tanabe O, Tanaka H, Kishimoto T (1988) Autocrine generation and requirement of BSF-2/IL-6 for human multiple myelomas. Nature 332:83–85

Leeman RJ, Lui VW, Grandis JR (2006) STAT3 as a therapeutic target in head and neck cancer. Expert Opin Biol Ther 6:231–241

Levy DE, Gilliland DG (2000) Divergent roles of STAT1 and STAT5 in malignancy as revealed by gene disruptions in mice. Oncogene 19:2505–2510

Liu X, Robinson GW, Wagner K-U, Garrett L, Wynshaw-Boris A, Hennighausen L (1997) Stat5a is mandatory for adult mammary gland development and lactogenesis. Genes Dev 11:179–186

Lui VW, Boehm AL, Koppikar P, Leeman RJ, Johnson D, Ogagan M, Childs E, Freilino M, Grandis JR (2007) Antiproliferative mechanisms of a transcription factor decoy targeting signal transducer and activator of transcription (STAT) 3: the role of STAT1. Mol Pharmacol 71:1435–1443

Lynch RA, Etchin J, Battle TE, Frank DA (2007) A small-molecule enhancer of signal transducer and activator of transcription 1 transcriptional activity accentuates the antiproliferative effects of IFN-gamma in human cancer cells. Cancer Res 67:1254–1261

Marotta LLC, Almendro V, Marusyk A, Shipitsin M, Schemme J, Walker SR, Bloushtain-Qimron N, Kim JJ, Choudhury SA, Maruyama R et al (2011) The JAK2/STAT3 signaling pathway is required for growth of CD44+CD24,Äì stem cell,Äìlike breast cancer cells in human tumors. J Clin Invest

Nelson EA, Walker SR, Kepich A, Gashin LB, Hideshima T, Ikeda H, Chauhan D, Anderson KC, Frank DA (2008) Nifuroxazide inhibits survival of multiple myeloma cells by directly inhibiting STAT3. Blood 112:5095–5102

Nelson EA, Sharma SV, Settleman J, Frank DA (2011a) A chemical biology approach to developing STAT inhibitors: molecular strategies for accelerating clinical translation. Oncotargets 2(6): 518–524

Nelson EA, Walker SR, Weisberg E, Bar-Natan M, Barrett R, Gashin LB, Terrell S, Klitgaard JL, Santo L, Addorio MR et al (2011b) The STAT5 inhibitor pimozide decreases survival of chronic myelogenous leukemia cells resistant to kinase inhibitors. Blood 117:3421–3429

Schwaller J, Parganas E, Wang D, Cain D, Aster JC, Williams IR, Lee C-K, Gerthner R, Kitamura T, Frantsve J et al (2000) Stat5 is essential for the myelo- and lymphoproliferative disease induced by TEL/JAK2. Mol Cell 6:693–704

Sen M, Tosca PJ, Zwayer C, Ryan MJ, Johnson JD, Knostman KA, Giclas PC, Peggins JO, Tomaszewski JE, McMurray TP, Li C, Leibowitz MS, Ferris RL, Gooding WE, Thomas SM, Johnson DE, Grandis JR (2009) Lack of toxicity of a STAT3 decoy oligonucleotide. Cancer Chemother Pharmacol 63:983–995

Socolovsky M, Nam H, Fleming MD, Haase VH, Brugnara C, Lodish HF (2001) Ineffective erythropoiesis in Stat5a(-/-)5b(-/-) mice due to decreased survival of early erythroblasts. Blood 98:3261–3273

Song H, Wang R, Wang S, Lin J (2005) A low-molecular-weight compound discovered through virtual database screening inhibits Stat3 function in breast cancer cells. Proc Natl Acad Sci USA 102:4700–4705

Szczepek AJ, Belch AR, Pilarski LM (2001) Expression of IL-6 and IL-6 receptors by circulating clonotypic B cells in multiple myeloma: potential for autocrine and paracrine networks. Exp Hematol 29:1076–1081

Trauger JW, Baird EE, Dervan PB (1996) Recognition of DNA by designed ligands at subnanomolar concentrations. Nature 382:559–561

Turkson J, Ryan D, Kim JS, Zhang Y, Chen Z, Haura E, Laudano A, Sebti S, Hamilton AD, Jove R (2001) Phosphotyrosyl peptides block Stat3-mediated DNA binding activity, gene regulation, and cell transformation. J Biol Chem 276:45443–45455

Turkson J, Kim JS, Zhang S, Yuan J, Huang M, Glenn M, Haura E, Sebti S, Hamilton AD, Jove R (2004) Novel peptidomimetic inhibitors of signal transducer and activator of transcription 3 dimerization and biological activity. Mol Cancer Ther 3:261–269

Uchiyama H, Barut BA, Mohrbacher AF, Chauhan D, Anderson KC (1993) Adhesion of human myeloma-derived cell lines to bone marrow stromal cells stimulates interleukin-6 secretion. Blood 82:3712–3720

Walker SR, Nelson EA, Frank DA (2007) STAT5 represses BCL6 expression by binding to a regulatory region frequently mutated in lymphomas. Oncogene 26:224–233

Walker SR, Nelson EA, Zou L, Chaudhury M, Signoretti S, Richardson A, Frank DA (2009) Reciprocal effects of STAT5 and STAT3 in breast cancer. Mol Cancer Res 7:966–976

Walker SR, Chaudhury M, Nelson EA, Frank DA (2010) Microtubule-targeted chemotherapeutic agents inhibit STAT3 signaling. Mol Pharmacology 78(5):903–908

Wang LH, Yang XY, Kirken RA, Resau JH, Farrar WL (2000) Targeted disruption of Stat6 DNA binding activity by an oligonucleotide decoy blocks IL-4-driven T_H2 cell response. Blood 95:1249–1257

Wang T, Niu G, Kortylewski M, Burdelya L, Shain K, Zhang S, Bhattacharya R, Gabrilovich D, Heller R, Coppola D et al (2004) Regulation of the innate and adaptive immune responses by Stat-3 signaling in tumor cells. Nat Med 10:48–54

Xi S, Gooding WE, Grandis JR (2005) In vivo antitumor efficacy of STAT3 blockade using a transcription factor decoy approach: implications for cancer therapy. Oncogene 24:970–979

Yang J, Stark GR (2008) Roles of unphosphorylated STATs in signaling. Cell Res 18:443–451

Chapter 8
Protein Therapeutics in Oncology

Michael J. Corbley[†]

Abbreviations

ADCC	Antibody-dependent cell-mediated cytotoxicity
ALL	Acute lymphoblastic leukemia
AML	Acute myeloid leukemia
APC	Antigen-presenting cell
CAF	Carcinoma-associated fibroblast
CDC	Complement-dependent cytotoxicity
CDR	Complementarity determining region
CEA	Carcinoma embryonic antigen
CLL	Chronic lymphocytic leukemia
CML	Chronic myelogenous leukemia
CSC	Cancer stem cell
CTL	Cytotoxic T lymphocyte
CTLA-4	Cytotoxic T-lymphocyte antigen-4
DR	Death receptor
EC	Endothelial cell
EGF	Epidermal growth factor
EMT	Epithelial-mesenchymal transition
FACS	Fluorescence-activated cell sorting
FGF	Fibroblast growth factor
Flt-3	fms-related tyrosine kinase-3
GM-CSF	Granulocyte macrophage-colony stimulating factor

M.J. Corbley (✉)
Deceased

D.A. Frank (ed.), *Signaling Pathways in Cancer Pathogenesis and Therapy*,
DOI 10.1007/978-1-4614-1216-8_8, © Springer Science+Business Media, LLC 2012

HBV	Hepatitis B virus
HGF	Hepatocyte growth factor
HPV	Human papilloma virus
IFN	Interferon
IGF	Insulin-like growth factor
IL	Interleukin
IR	Insulin receptor
ISG	IFN-stimulated genes
LAK	Lymphokine-activated killer cell
LPS	Lipopolysaccharide
LTBR	Lymphotoxin beta receptor
mAb	Monoclonal antibody
MIF	Macrophage migration inhibitory factor
MMP	Matrix metalloproteinase
MOA	Mechanism-of-action
MUC1	Mucin-1
NHL	Non-Hodgkins lymphoma
NK	Natural killer cell
NSCLC	Nonsmall cell lung cancer
PSA	Prostate-specific antigen
RA	Rheumatoid arthritis
RANK	Receptor activator of NFkB
RCC	Renal cell carcinoma
RTK	Receptor tyrosine kinase
S1P	Sphingosine-1-phosphate
scFv	Single-chain variable fragment antibody
SDF-1	Stromal cell-derived growth factor-1
TAM	Tumor-associated macrophage
TGF	Transforming growth factor
TLR	Toll-like receptor
TME	Tumor microenvironment
TNF	Tumor necrosis factor
TRAIL	TNF-related apoptosis-inducing ligand
Treg	Regulatory T cell
VDA	Vascular disrupting agent
VEGF	Vascular endothelial growth factor

8.1   Introduction

Protein therapeutics have become an integral component of the battle against cancer. This review seeks to provide a broad overview of the field, highlighting recent advances and describing the direction of ongoing research. The emphasis is on the mechanism-of-action of different classes of protein therapeutics and on the roles that their targets play in various sectors of cancer biology, rather than on specific types of cancer. Clinical-stage examples are provided for each class and novel protein-based therapeutic concepts are noted. Recent relevant advances in antibody engineering are also covered as a separate topic.

Most protein therapeutics are intended to act extracellularly, either on the tumor cells themselves or on other cells that interact with the tumor; they thereby complement the many small-molecule chemotherapies that usually act intracellularly. While this review covers only proteins, the reader should keep in mind that in some cases (e.g., the transmembrane receptor kinases), small-molecule or other therapeutic modalities may also be in development for the same target.

8.2 Anti-Tumor Strategies

Most antibody or recombinant protein therapeutics are designed with a particular anti-tumor strategy or mechanism-of-action (MOA) in mind. These strategies include:

1. Killing tumor cells directly by antibody-dependent cell-mediated cytotoxicity (ADCC) and complement-dependent cytotoxicity (CDC) resulting from effector activity of the Fc portion of an antibody.
2. Causing tumor cells to undergo apoptosis as a result of the perturbation caused by binding of an antibody to their cell-surface antigens.
3. Delivering a toxic payload to the tumor by means of an antibody or fusion protein.
4. Interrupting cell-proliferation pathways by preventing activation of transmembrane receptor enzymes, such as receptor tyrosine kinases. This may involve blocking of ligand binding or inhibition of a physical event, such as receptor dimerization. Antibodies and decoy receptors can be used. Therapeutics in this category may be cytostatic rather than cytotoxic, since they prevent proliferation, although apoptosis can still occur.
5. Augmenting the anti-tumor immune response or suppressing immune tolerance, such that the immune system destroys the tumor. Interferons, cytokines, vaccines, and adjuvants fall into this category.
6. Activating natural apoptotic pathways. Agonist antibodies to the TNF family are an example.
7. Altering the tumor stroma rather than the tumor itself, in such a way as to prevent the stroma from supporting the tumor. Angiogenesis inhibitors fall into this category.

The protein therapeutics mentioned in this article are listed in Table 8.1, together with their targets and their developer or current sponsor.

Table 8.1 Protein therapeutics referred to in this chapter. An asterisk indicates that the therapeutic is approved

Name	Synonym	Type	Target	Sponsor
*Rituximab	Rituxan	Ab	CD20	Biogen Idec, Roche
Ofatumumab		Ab	CD20	GenMab, GSK
*Alemtuzumab	Campath	Ab	CD52	Genzyme
Epratuzumab	IMMU-103	Ab	CD22	Immunomedics
Galiximab	IDEC-114	Ab	CD80	Biogen Idec
Lumiliximab	IDEC-152	Ab	CD23	Biogen Idec
SGN-30		Ab	CD30	Seattle Genetics
Lintuzumab	SGN-33	Ab	CD33	Seattle Genetics
*Cetuximab	Erbitux	Ab	EGFR	ImClone/Eli Lilly
*Panitumumab	Vectibix	Ab	EGFR	Amgen
*Trastuzumab	Herceptin	Ab	HER2	Genentech/Roche
Pertuzumab	Omnitarg	Ab	HER2	Roche, Chugai
Rilotumumab	AMG 102	Ab	HGF	Amgen
NK4		HGF fragment	c-Met	Kringle, Osaka U.
MetMab		monovalent Fab	c-Met	Genentech/Roche
Figitumumab	CP-751871	Ab	IGF-1R	Pfizer
MK-0646		Ab	IGF-1R	Merck
AMG-479		Ab	IGF-1R	Amgen, Takeda
BIIB-022		Ab	IGF-1R	Biogen Idec
Robatumumab	SCH-717454	Ab	IGF-1R	Schering-Plough
Cixutumumab	IMC-A12	Ab	IGF-1R	ImClone/Lilly
RG-1507		Ab	IGF-1R	GenMab, Roche
*Ibritumomab tiuxetan	Zevalin	radiolabeled Ab	CD20	Spectrum Pharma
*131I-Tositumomab	Bexxar	radiolabeled Ab	CD20	GlaxoSmithKline
90Y-epratuzumab	IMMU-102	radiolabeled Ab	CD22	Immunomedics
*Gemtuzumab ozogamicin	Mylotarg	drug-Ab	CD33	Wyeth, Celltech
Trastuzumab-DM1		toxin-Ab	HER2	Genentech/Roche
*IFN-α2	Roferon-A	interferon		Roche, Takeda

*IFN-α2, pegylated	Pegasys	interferon		Roche
*IL2	Aldesleukin	cytokine		Chiron/Novartis
*Denileukin diftitox	Ontak	toxin-IL2 fusion		Eisai
Tucotuzumab celmoleukin	EMD-273066	IL2 fusion	EpCAM	Merck Serono
L19-IL-2		scFv-IL2 fusion	oncofetal FN ED-B	Bayer Schering
EMD-273063		Ab-IL2 fusion	GD2	Merck Serono
IL-7	CYT-107	cytokine		Cytheris
IL-12		cytokine		NIH
*GM-CSF	sargramostim	cytokine		Genzyme
Sipuleucel-T	Provenge	vaccine		Dendreon
Ipilimumab	MDX-010	Ab	CTLA-4	Bristol-Meyers Squibb
*TNFα	tasonermin	cytokine		Genentech, Boehringer Ing
golnerminogene pradenovec	TNFerade	TNF gene therapy		GenVec
L19-TNFα		scFv-TNFα fusion	oncofetal FN ED-B	Bayer Schering
RhApo2L/TRAIL	dulanermin	recomb. ligand	DR4, DR5	Amgen, Roche
Mapatumumab	HGS-ETR1	agonist Ab	DR4	Human Genome Sciences
Apomab	RG-7425	agonist Ab	DR5	Genentech/Roche
Lexatumumab	HGS-ETR2	agonist Ab	DR5	Human Genome Sciences
Conatumumab	AMG-655	agonist Ab	DR5	Amgen, Takeda
Tigatuzumab	CS-1008	agonist Ab	DR5	Daiichi Sankyo
Dacetuzumab	SGN-40	Ab	CD40L	Seattle Genetics
Lucatumumab	CHIR 12.12	Ab	CD40L	Novartis, XOMA
*Bevacizumab	Avastin	Ab	VEGF	Genentech/Roche
Ramucirumab	IMC-1121B	Ab	KDR/VEGFR2	ImClone/Eli Lilly
Aflibercept	VEGF trap	decoy receptor	VGEF	Regeneron, Sanofi-Aventis

(continued)

Table 8.1 (continued)

Name	Synonym	Type	Target	Sponsor
IMC-EB10		Ab	Flt-3	ImClone/Eli Lilly
Sonepcizumab	Sphingomab	Ab	S1P	Lpath, Merck Serono
AMG-386		peptibody	angiopoietins	Amgen, Takeda
CT-322		adnectin	VEGF	Adnexus/BMS
Endostatin		recomb. protein		Children's Hospital, Boston
Angiostatin		recomb. protein		Children's Hospital, Boston
NGR-hTNF	Arenegyr	peptide-TNF fusion	CD13	MolMed
Bavituximab		Ab	Phosphatidylserine	Peregrine
Sibrotuzumab	BIBH-1	Ab	FAP	Ludwig Institute
Denosumab	AGM-162	Ab	RANKL	Amgen
DX-2400		Ab	MMP-14	Dyax
GC-1008		Ab	TGFβ	Genzyme
Intetumumab	CNTO-95	Ab	$\alpha v \beta 3$, $\alpha v \beta 5$	Centocor/J&J
Etaracizumab	MEDI-522	Ab	$\alpha v \beta 3$	MedImmune/ AstraZeneca
Volociximab	M-200	Ab	$\alpha 5 \beta 1$	Facet, Biogen Idec
PF-04605412		Ab	$\alpha 5 \beta 1$	Pfizer
*Natalizumab	Tsyabri	Ab	$\alpha 4 \beta 1$	Elan, Biogen Idec
Catumaxomab	removab	bispecific Ab	EpCAM/CD3	Fresenius, TRION
Edrecolomab			EpCAM	Centocor, J&J
Adecatumumab	MT-201	Ab	EpCAM	Centocor, J&J
*L-asparaginase	Oncaspar	enzyme	Asparagine	Enzon
Ranpirnase	Onconase	enzyme	RNA	Alfacell
BH3-stapled peptides	BIM-SAHB	peptide	BAX	Aileron
AS-1411		aptamer	Nucleolin	Antisoma
Blinatumomab	MT-103	bispecific Ab	CD19/CD3	Micromet

8.3 Antibodies Against Cell-Surface Antigens and Receptors

The workhorse of oncology protein therapeutics is the monoclonal antibody (mAb), directed against an antigen on the surface of the tumor cell or against a ligand that interacts with a tumor cell-surface receptor (Fanale and Younes 2007; Castillo et al. 2008). Antibodies owe their success to their exquisite specificity, long half-life, and a relative lack of side effects when compared to traditional chemotherapeutics. Table 8.2 lists the first nine antibodies approved for oncology, along with their mechanisms-of-action. The MOAs are as diverse as the antibodies themselves and in some cases multiple MOAs appear to be at work.

8.3.1 Cell-Surface Antigens

Rituximab was the first mAb cancer drug, approved in 1997 for non-Hodgkin's lymphoma (NHL). It has been used with dramatic success in diffuse large B-cell lymphoma and follicular lymphoma and is widely prescribed for chronic lymphocytic leukemia (CLL) as well. The antibody recognizes CD20, an antigen of unknown function on the surface of most B-cells. When bound to CD20, rituximab stimulates ADCC and CDC according to design, although recent work suggests that it also directly induces apoptosis in malignant B-cells. Interestingly, the epitope recognized by the antibody appears to be important: Ofatumumab is a new anti-CD20 monoclonal in late-stage trials. It binds a different epitope from rituximab, resulting in significantly enhanced CDC in cell culture and in *ex vivo* CLL patient samples, and it has shown responses in CLL patients who have relapsed on rituximab (Pawluczkowycz et al. 2009).

Table 8.2 Cellular mechanisms-of-action of current oncology antibody drugs

Antibody	Trade name	Mechanisms-of-action
Rituximab	Rituxan	ADCC, CDC
		Direct induction of apoptosis
Alemtuzumab	Campath	ADCC, CDC
Traztuzumab	Herceptin	Inhibition of receptor dimerization
		ADCC
		Inhibition of shedding of extracellular domain
Cetuximab	Erbitux	Competitive inhibition of ligand binding
		Receptor down-regulation
		ADCC
Panitumumab	Vectibix	Competitive inhibition of ligand binding
		Receptor down-regulation
		ADCC
Bevacizumab	Avastin	Inhibition of ligand binding to receptor
Gemtuzumab	Mylotarg	Internalization of toxic drug payload
Ibritumomab	Zevalin	Radioactive payload killing of tumor cells
Tositumomab	Bexxar	Radioactive payload killing of tumor cells

Building on the success of rituximab, a popular approach for hematological malignancies has been to generate antibodies against other cell-surface antigens (Castillo et al. 2008), without regard to their function, assuming *a priori* that these antibodies will elicit direct cell-killing by ADCC or CDC. For example, Alemtuzumab is an approved antibody against CD52, an antigen found on lymphocytes and various myeloid cells, as well as CLL cells. The drug is approved for fludarabine-refractory CLL and is capable of eliminating minimal residual disease from bone marrow. The MOA is believed to be ADCC or CDC and is not directly related to the function of CD52 (Mould et al. 2007).

Further research on a promising antibody may subsequently show that efficacy depends on an antigen-specific MOA, in addition to or instead of ADCC/CDC. For example, CD22 is involved in cellular adhesion and B-cell targeting. Epratuzumab (Leonard and Goldenberg 2007) is an anti-CD22 monoclonal in phase III trials for NHL, as well as various autoimmune diseases. This antibody is interesting in that it is rapidly internalized along with the CD22 antigen, suggesting that the MOA is directly related to CD22 function in activated B-cells. Galiximab (Leonard et al. 2007) is a primatized mAb that recognizes CD80 and is currently in Phase III trials for NHL. CD80 is found on malignant B-cells, the Reed-Sternberg cells in Hodgkins lymphoma, and antigen-presenting cells (APCs). It functions as the receptor for CD28, a costimulatory antigen on T-cells that is involved in T cell activation and inflammation. The MOA of galiximab was initially expected to be ADCC and this probably occurs; however, significant responses have occurred in patients with a delay of 6–12 months, when serum levels of galiximab are undetectable, suggesting that an MOA involving T cell-mediated immune system activation is also at work. Importantly, NHL patients treated with both galiximab and rituximab appear to fare better than those with rituximab alone, confirming preclinical results and suggesting that combination therapy with the various antibodies will be beneficial and perhaps finally curative. Antibodies against other CD antigens, including CD23 (lumiliximab), CD30 (SGN-30), and CD33 (lintuzumab), are in similar midstage trials for hematological malignancies.

8.3.2 *ErbB Family*

For solid tumors, new cell-surface antigens have more frequently been selected based on a known cell-proliferation signaling pathway that can be interrupted. This approach is exemplified by antibodies to the ErbB family (Hynes and Lane 2005), which is composed of four transmembrane receptors (ERBB1-4 or HER1-4). ERBB1, 3 and 4 are activated by ligand binding. ERBB1, 2 and 4 have intracellular kinase domains. In all cases, activation leads to receptor homo- or heterodimerization, kinase activation, and downstream signaling through the Ras/Raf/ERK proliferation pathway and the phosphatidylinositol 3-kinase (PI3K)/AKT survival pathway. For ERBB1, the epidermal growth factor receptor (EGFR), the antibodies cetuximab and panitumumab act as competitive inhibitors of EGF binding, thereby preventing dimerization (Ciardiello and Tortora 2008). These antibodies are approved for metastatic colorectal cancer and some cases of squamous cell head-and-neck cancer, and are in other clinical trials. ERBB2 (known more commonly as HER2 or HER2/neu) does not have a ligand; instead dimerization is induced by overexpression, mutation, or cleavage of the extracellular domain. The antibody trastuzumab appears to directly block this dimerization. However, ADCC must also play an important role in the MOA, since mice lacking receptors for the ADCC-inducing Fc domain of the antibody fail to respond to trastuzumab. Trastuzumab is approved for and clearly extends survival in cases of HER2-positive breast cancers, HER2 overexpression being a poor prognostic indicator (Hudis 2007).

While experience has been gained with these marketed therapeutics, it has come to be appreciated that the other family members ERBB3 (HER3) and ERBB4 (HER4) are also important in cancer (Baselga and Swain 2009). With six binding sites for PI3K, HER3 is especially active in promoting cell survival. Moreover, all pair-wise combinations of the four family members can occur as active heterodimers, perhaps limiting the range of the approved therapeutics. Therefore, the next generation of antibodies against this family are particularly intended to prevent heterodimerization, especially with HER3. Pertuzumab is in phase III trials for breast and ovarian cancer. It binds HER2 at an epitope distinct from the trastuzumab epitope and inhibits HER2 dimerization with the other family members. In preclinical models, it inhibited cell growth stimulated by heregulin, the ligand for HER3. Thus, it may prove efficacious in a wider range of cancers in which HER2 is active without being overexpressed, whereas trastuzumab only works in the 30% of breast cancers that overexpress HER2. HER3-specific and new EGFR-specific antibodies are in earlier development. Preliminary data suggest that the new antibodies may be synergistic with cetuximab, panitumumab, and trastuzumab.

8.4 Receptor Tyrosine Kinases

As a result of the successful introduction of the ERBB family therapeutics, nearly all transmembrane receptor tyrosine kinase (RTK) pathways have become potential targets in oncology. In many cases, the first strategy is to develop a small-molecule inhibitor of the kinase activity and the spectacular efficacy of imatinib in chronic myelogenous leukemia has justified this approach. However, in other cases, there are clear advantages to an extracellular protein-based approach as evidenced by the dimerization discussion above. Two of the most exciting examples are the c-Met- and IGF-1R-signaling axes.

8.4.1 c-Met

This proto-oncogene was originally isolated by chemical transformation of a nontumorigenic osteogenic sarcoma cell line. c-Met is the receptor for hepatocyte growth factor/scatter factor (HGF). HGF is a pleiotrophic cytokine and morphogenic factor secreted by mesenchymal cells; it promotes migration and proliferation of epithelial and endothelial cells, and it functions in the three-dimensional organization of glands and tubules during normal development. HGF binding to c-Met causes receptor dimerization and c-terminal tyrosine phosphorylation, resulting in kinase activation. In a range of solid tumors, in addition to stimulating proliferation and survival, the c-Met/HGF pathway appears to be integrally involved in invasion and metastasis, and it is also involved in angiogenesis. Constitutive activation occurs through amplification of the c-Met gene locus, receptor overexpression leading to ligand-independent dimerization, or point mutations in the kinase domain. Overexpression or aberrant expression of HGF has also been seen.

The importance of the HGF/c-Met pathway has led to a plethora of inhibition strategies (Peruzzi and Bottaro 2006; Toschi and Janne 2008). Several mAbs against the ligand HGF are in clinical trials, including rilotumumab (AMG 102), currently in phase II for renal cell carcinoma and glioma. Promising preclinical results have been seen with HGF mimetics, including NK4, a 447-amino acid internal fragment of HGF, which acts as a competitive antagonist to the ligand. Antibodies against the extracellular domain of the c-Met receptor have also been developed and shown to block HGF binding. MetMab is the most advanced, having progressed to phase II for nonsmall cell lung cancer (NSCLC) (Jin et al. 2008). Another anti-Met antibody caused receptor down-regulation through proteolytic cleavage, resulting in "shedding" of a soluble fragment of the extracellular domain (Petrelli et al. 2006). This study paralleled the development of a "decoy" c-Met ectodomain protein that inhibited HGF binding and receptor dimerization, leading to xenograft regression and metastasis inhibition in preclinical studies (Michieli et al. 2004). In addition to the protein therapeutics, small-molecule inhibitors of the Met kinase are also in development.

8.4.2 IGF-IR

The insulin-like growth factor receptor-1 (IGF-1R) may be the most-targeted cell-surface receptor in oncology, based on 20 years of laboratory and epidemiological evidence linking the pathway to cancer (Pollak 2008). A member of the insulin receptor (IR) superfamily, IGF-IR is part of a complex signaling network, which includes two ligands (IGF-1 and IGF-2), six IGF-binding proteins and two receptors (IGF-IR and IGF-IIR, which lacks an intracellular signaling domain). IGF-1 and −2 are produced by the liver in response to growth hormone, or by tumor cells. The IGFs stimulate dimerization of IGF-IR, as insulin stimulates dimerization of IR, resulting in autophosphorylation and signaling. Both IGF-1 and insulin can stimulate heterodimerization of IGF-IR with IR. The downstream effector pathways include the PI3K/AKT/mTOR survival pathway and the Raf/MEK/Erk proliferative pathway. The AKT/mTOR pathway regulates energy balance within the cell and is now central to a new paradigm in which cancer is seen as a disease of energy metabolism. Thus, the IGF-IR pathway serves not only in the maintenance of the transformed phenotype, but also activates prosurvival mechanisms and suppresses apoptosis in cancer cells. Clinically, IGF-IR has been linked to development of drug resistance to Her-2 inhibitors in breast cancer, to EGFR inhibitors in lung cancer and to mTOR inhibitors (e.g., rapamycin) in various solid tumors. Aberrations in nearly every network component have been observed in tumors, including overexpression, amplification or loss of imprinting of the ligands or receptors, and loss of IGF-IIR, which acts as a negative regulator of the pathway.

One aspect that sets insulin biology apart from growth factors such as EGF is that insulin and the IGFs regulate the physiology of the entire organism; IGF-IR and IR are significantly expressed in nearly all normal tissues. Hence, the potential for unwanted toxicity is unusually high. The ATP-binding sites of the IGF-IR and IR share 95% homology, which has made it difficult to develop small molecules that are specific for IGF-IR. Antibodies against IGF-IR are therefore seen as the more viable approach. At least seven mAbs (figitumumab (CP-751871), MK-0646, AMG-479, BIIB-022, robatumumab (SCH-717454), cixutumumab (IMC-A12) and RG-1507) of varying composition are in midstage clinical development in various cancers and encouraging results have been observed, both as monotherapies and in combination with chemotherapy or radiation (Pollak 2008; Rodon et al. 2008). MOAs range the gamut from ADCC to inhibition of ligand binding to receptor down-regulation, depending on the antibody structure. Variable toxicity has also been reported, with anticipated hyperglycemia being the most common. Meanwhile, other protein strategies in preclinical development include soluble IGF-IR decoys to sop up the IGFs, antibodies against the IGFs and recombinant forms of the IGF-binding proteins.

8.5 Immunoconjugates

A different approach with antibodies has been to use them as vehicles to deliver a cytotoxic agent to the tumor, making use of their high specificity and tight binding. The first approved immunoconjugates employed a radionuclide conjugated to an anti-CD20 antibody, yielding ibritumomab tiuxetan, which carries yttrium-90, and 131I-tositumomab, which carries iodine-131 (Ricart and Tolcher 2007). These drugs are indicated for the treatment of relapsed or refractory low-grade, follicular, or transformed B-cell NHL or for follicular NHL refractory to rituximab. They deliver a high dose of radioactivity directly to tumor cells expressing the antigen and have the additional benefit of bystander killing of neighboring tumor cells and tumor-supporting stromal cells. While these therapeutics have demonstrated a prolonged survival benefit relative to the naked antibody drug, logistical problems have prevented them from being widely prescribed. They must be administered in a nuclear medicine department rather than by an oncologist; they require the stringent precautions associated with radioactive materials and they have not been readily reimbursable.

Despite the difficulties, however, research has continued. Several new radioimmunoconjugates have reached clinical trials, including a 90Y-conjugated epratuzumab (anti-CD22, mentioned above), while others have been discontinued. The list of useful radionuclides has been expanded to include beta emitters other than 131I and 90Y, and shorter-pathlength alpha emitters. Nonetheless, progress has been slow, particularly in solid tumors, which are less sensitive to radiation and where tumor penetration is difficult. It may turn out that radioimmunotherapies will achieve their usefulness in elimination of minimal residual disease or of undetectable micrometastases.

Instead, future immunoconjugates will likely involve payloads consisting of toxins or drugs, thereby eliminating the difficulties with radioactivity (Wu and Senter 2005; Ricart and Tolcher 2007). Early results with protein toxins such as ricin and diphtheria toxin were disappointing, due to severe liver and kidney toxicity, vascular leak syndrome, and rapid development of neutralizing antibodies against the toxin itself. Conjugates with drugs such as the standard chemotherapeutic doxorubicin also proved unsuccessful. However, gemtuzumab ozogamicin, which consists of the cytotoxic drug N-acetyl-gamma calicheamicin connected by an acid-labile hydrazone linker to an anti-CD33 mAb, was approved in 2000 for acute myeloid leukemia (AML). New payloads are still being developed, including the natural product maytansine, a very potent anti-tubulin agent.

Meanwhile, a great deal of work has gone into improving the structure of immunoconjugates (Senter 2009). In particular, linkers have been modified to be more stable in circulation and to release the payload immediately upon entry into the cell. In the case of maytansine, the payload is bound by disulfide linkage to the antibody. Release of the active maytansine occurs rapidly only in the high reducing environment inside the cell. Clinically, the leading example is trastuzumab-DM1, currently in phase III trials for HER-2-positive metastatic breast cancer after achieving significant response rates in Phase II (Lewis Phillips et al. 2008). An even more potent payload, DM4, is the current choice for newer antibodies.

Further progress has been made in immunoconjugate engineering (Wu and Senter 2005). For example, single-chain antibodies containing only the recognition domains (scFv) have been utilized to reduce the time in circulation from weeks to hours, thereby reducing the accumulation of normal tissue toxicity. Site-specific conjugation on the antibody is now possible, allowing for consistent control of the number and location of payload molecules. Finally, an elaborate approach to reduce toxicity known as pretargeting involves an initial treatment with the antibody conjugated to a molecule such as streptavidin, a pause to allow unbound antibody to clear from the body, and then a second treatment with a radionuclide-conjugated biotin. The hope is that the tight biotin-streptavidin association will result in enhanced delivery of the radionuclide to the tumor, while avoiding normal tissue.

In determining at the outset whether to pursue a payload approach, factors that may predict success include antigen/antibody complexes that are rapidly internalized (e.g., CD22), antigens that are highly specific to tumor tissue, and antibody subtypes that only poorly facilitate effector cell-dependent killing.

8.6 Interferons and Cytokines and Cancer Vaccines

The theory of immunosurveillance (Smyth et al. 2001) implies that tumors must evade or actively suppress the immune system in order to survive and grow. Hence, the field of cancer immunotherapy has developed with the intention of revitalizing the immune reaction to the tumor. From the protein perspective, strategies have included immune stimulation with cytokines and interferons, interference with immunosuppression by means of antibodies, and elicitation of *de novo* immune responses by immunization with tumor antigens.

8.6.1 IFNα

The interferons (IFN) are a family of 20 cytokines that generate resistance to viruses and pathogens, stimulate cells of the innate immune system, enhance antigen presentation to T cells, and help to eliminate cancer cells (Borden et al. 2007). The α-IFNs are widely used in treating viral diseases and hepatitis B and C, while the β-IFNs have been effective in multiple sclerosis. In oncology, IFN-α2 was the first significant cancer therapy to emerge from recombinant DNA technology, being approved for hairy cell leukemia in 1986. Since then, a wide range of trials have established the cancers for which the IFN-α2 is effective (chronic myelogenous leukemia (CML), lymphomas, melanoma, renal and bladder carcinoma) or not effective (lung, colon, breast carcinomas) (Goldstein and Laszlo 1988). IFN appears to act both directly on tumor cells to induce apoptosis and indirectly by stimulating natural killer (NK) cells, lymphokine-activated killer (LAK) cells, and other components of the immune system. Addition of polyethylene glycol side chains improved the IFNs by increasing serum half-life, allowing less frequent dosing. Preclinical experiments suggest that IFN-β is a more potent anti-cancer agent than IFN-α, but translation to the clinic has lagged. IFN-γ is potentially the most potent inducer of the immune response, eliciting a Th1 polarization, but its use in oncology has been unsuccessful. Meanwhile, even the use of IFN-α has been supplanted by better therapeutics in some cases, as with the use of the abl small-molecule kinase inhibitor imatinib in CML. Current clinical research is primarily directed towards determining where synergies can be achieved by combining IFN therapy with other therapies, for example, with bevacizumab to enhance anti-angiogenic activity or with toll-like receptor (TLR) inducers to enhance innate immune response (Borden 2005). Preclinically, research is focusing on direct activation of the hundreds of IFN-stimulated genes (ISG). Establishing correct dosing is still a matter of interest, because direct cytotoxicity occurs at high doses while low doses are sufficient for immune stimulation.

8.6.2 Cytokines

The most notable cytokine currently used in oncology is interleukin-2 (IL-2). First identified in 1965, IL-2 is a master regulator of immune response, driving the proliferation of antigen-activated T cells. Acting through the multisubunit IL-2 receptor (IL-2R), it controls the development of memory T cells, the maturation of cytotoxic T lymphocytes (CTLs) as well as immune-suppressing regulatory T cells (Tregs), and the production of antibodies by B cells. Recombinant human IL-2, produced in E. coli (aldesleukin), is approved for renal cell carcinoma (RCC) and malignant melanoma (Atkins 2002). High-dose therapy has generated remarkable results in a minority of patients. The MOA does not involve the tumor directly; rather, IL-2 stimulates T helper cells to secrete other cytokines, including TNF and IFN-γ, thereby increasing the immune attack on the tumor. It also induces the proliferation of CTLs and the activation of NK cells. Unfortunately, the therapeutic window is narrow and high-dose IL-2 therapy has proven to be quite toxic, due to its pleiotrophic action on all aspects of the immune system and the associated cytokine storm, resulting in side-effects in multiple organ systems.

Due to the obvious therapeutic value of IL-2 and its potential in a variety of tumors in addition to RCC and melanoma, the improvement of IL-2 therapy is one of the most active areas of research. Various engineered mutations ameliorate toxicity by decreasing NK activation relative to T cells, by reducing the secondary cytokine release, or by suppressing capillary leak syndrome. Liposomal, controlled-release and inhaled versions of IL-2 are all in active clinical trials. Fusion proteins incorporating IL-2 and another component have also made headway (Khawli et al. 2008). The other component may be an antibody against a cell-surface tumor antigen or it may be another protein that homes to the tumor. In either case, the idea is to localize IL-2 to the tumor to enhance the IL-2-associated immune response, while sparing normal tissues. Denileukin diftitox, in which IL-2 is fused to a diphtheria toxin fragment, was the first successful example, being approved in 1999 for persistent or recurrent cutaneous T-cell lymphoma. Current midstage clinical candidates include fusions of IL-2 with an anti-EpCAM antibody (tucotuzumab celmoleukin), with an anti-oncofetal fibronectin ED-B domain antibody (L19-IL-2), both for solid tumors, and with an anti-GD2 antibody (EMD-273063), for GD2-expressing neuroblastoma and melanoma.

Finally, IL-2 administration in conjunction with autologous hematopoietic cell transplant for leukemia failed to increase survival. However, low-dose IL-2 is now being used as an adjuvant to enhance anti-tumor vaccines. In some instances, a functional IL-2 gene has even been incorporated into the killed tumor cells or viral vectors of the vaccine or into reinfused TIL cells, to insure IL-2 expression at the site of action.

8.6.3 Other Interleukins

Meanwhile, the search is on for a less pleiotrophic cytokine with the anti-tumor potency of IL-2, but fewer systemic side-effects. In 2007, the National Cancer Institute Workshop issued its "top-20 list" of desirable new agents for cancer immunotherapy (http://web.ncifcrf.gov/research/brb/workshops/NCI%20Immunotherapy %20Workshop%207-12-07.pdf). IL-15, IL-12, and IL-7 occupied positions 1, 3, and 5, respectively. IL-15 and IL-7 share the gc receptor subunit with IL-2 and thus stimulate

T-cell maturation. Despite some preclinical successes, recombinant IL-7 is only in phase I and no IL-15 trials appear to be running. IL-12 may be the heir-apparent to IL-2. It coordinates the innate and adaptive immune responses, causing the polarization of T cells to a Th1 phenotype, the secretion of IFN-γ, and the stimulation of CD8 and NK cell cytotoxicity. Additional anti-tumor activities of IL-12 include increased expression of class I and class II MHC molecules and suppression of angiogenesis inducers. Recombinant IL-12 is currently being evaluated for a variety of cancers, alone and in combination with IL-2, as well as in conjunction with dendritic cell-based vaccines. Vectors expressing IL-12 are also under investigation. Preclinical research suggests that other interleukins may have anti-tumor efficacy as well (Margolin 2008).

Still, results with other cytokines in monotherapy have been disappointing. It is now generally agreed that combinations of cytokines at moderate doses are most likely to provide a therapeutic window and significant clinical benefit, but many years of clinical trials will be required, since all but IL-2 are still experimental. IL-12 is considered a top candidate for combination cytokine therapy (Weiss et al. 2007). To the extent that cytokine therapy turns out to provide long-term cures, long-term immunological side effects may still be a concern: these cytokines all stimulate the immune system and antibodies against all of them are being developed as immunosuppressives for autoimmune indications.

8.6.4 GM-CSF

Granulocyte macrophage-colony stimulating factor (GM-CSF) is a different cytokine that stimulates the growth and development of precursors of granulocytes and macrophages, rather than lymphocytes. Recombinant GM-CSF produced in yeast (sargramostim) is a vital drug for the treatment of chemotherapy- or radiotherapy-induced neutropenia and other cytopenias (Waller 2007). It does not have anti-cancer applications on its own. However, like IL-12, it is a potent activator of antigen-presenting dendritic cells. In one animal study of many cytokines and other secreted or surface molecules, GM-CSF proved to be the most potent immunostimulatory molecule in combination with experimental vaccines (Dranoff 2002). Thus, on account of its ready availability and long track record, both GM-CSF itself and expression constructs incorporating its gene have been widely tested as adjuvants to cancer vaccines. In addition, it is proving to synergize with drugs such as rituximab in B-cell lymphomas, due to its stimulation of effector cells for ADCC, especially including macrophages, which some consider to be of predominant importance in rituximab therapy (Schuster et al. 2008).

8.6.5 Cancer Vaccines

Cancer vaccines are frequently based on recombinant proteins. They can be divided into two categories: preventive and treatment vaccines (http://www.cancer.gov/cancertopics/factsheet/cancervaccine). Two cancer preventive vaccines are marketed in the U.S. A vaccine against hepatitis B virus (HBV) is intended to prevent liver damage, including hepatocellular carcinoma. This vaccine consists of hollow particles composed of the viral surface antigen HBVsAg. A vaccine against human papillomavirus (HPV) is intended to prevent cervical cancer. It consists of virus-like particles generated from the surface antigens of HPV types 6, 11, 16, and 18. Both vaccines are quite effective, although the HPV vaccine only protects against 70% of cervical cancer cases, the other 30% being caused by at least 17 other HPV types.

Cancer treatment vaccines are intended to suppress cancers which have already appeared or to prevent the reappearance of cancers in remission. Many approaches have been taken and vaccines are being tested in nearly every type of cancer (Begley and Ribas 2008; Copier et al. 2009). Some are based on known tumor antigens, including carcinoma embryonic antigen (CEA), mucin-1 (MUC1), and prostate-specific antigen (PSA). Others are based on the entire antigen profile of a specific tumor. They come as proteins, as nucleic acids expressing the proteins, as killed tumor cells, and as dendritic or other antigen-presenting cells, which may have been engineered to produce a particular protein. The scope is too broad to be covered here. Thus far, no cancer treatment vaccines have been approved and many have failed. However, hopes have been raised by recently reported results of a Phase III trial of sipuleucel-T in metastatic hormone-resistant prostate cancer that showed a statistically significant 4.1-month increase in median survival and a 38% increase in 3-year survival compared to placebo. The autologous therapy involves incubating a patient's own antigen-presenting cells with a fusion protein consisting of the stimulating antigen prostatic acid phosphatase and the cytokine GM-SCF.

8.6.6 Ipilimumab

Cytotoxic T-lymphocyte antigen-4 (CTLA-4) is a cell-surface antigen expressed by activated T cells. It competes with CD28, the primary costimulatory molecule on T cells, for binding to CD80 and CD86 on antigen-presenting cells. CTLA-4 thereby suppresses an immune response. Absence or mutation of CTLA-4 leads to severe lymphoproliferative disorders. Conversely, in some cancers, CTLA-4 appears to be responsible for suppression of the anti-tumor response. Ipilimumab is an anti-CTLA-4 antibody that blocks the association with CD80 and CD86, thereby relieving this suppression. It is currently in late-stage trials as monotherapy and in association with tumor vaccines. Earlier trials have apparently shown impressive durable responses in a minority of metastatic melanoma patients and results of the registrational trials are eagerly awaited to determine whether response rate will be high enough to warrant approval. It is notable that some patients initially progressed on therapy and were considered nonresponders and then experienced significant tumor regression after many months. This delayed response illustrates one of the conundra associated with currently accepted solid tumor response criteria; yet, it is the type of response that might be expected from a therapeutic which promotes an immune response, as opposed to directly acting on the tumor (Weber 2008).

8.7 The TNF Superfamily

So far, the discussion has dealt with inhibition or elimination of targets involved in initiation of oncogenesis or maintenance of the transformed phenotype. Another promising strategy involves the harnessing of proteins whose normal function is to protect against cancer or to naturally induce apoptosis. Examples include the tumor necrosis factor (TNF) superfamily.

The TNF superfamily consists of 18 ligands and 28 receptors. The ligands exist as homotrimers and are usually membrane-associated. The receptors are all trans-membrane and generally multimerize upon ligand binding. Some have an internal death domain; none possess enzymatic activity, but rather elicit intracellular signaling after association of a complex of proteins with the intracellular tail. While the signaling pathways vary, the end result relevant to oncology is the activation of caspases that kill the cell. In the classical model of TNF activity, the TNF receptor is linked to the extrinsic apoptosis pathway, but in some instances the cytochrome c-enabled intrinsic pathway is also activated. Initiation of killing in one cell can occur through engagement of a TNF-family receptor by a soluble ligand or by a membrane-bound ligand on another cell.

8.7.1 TNFα

TNFα has been known as a tumor-necrotizing substance related to infection for more than a century; it was originally isolated as the cytotoxic factor present in mouse serum after stimulation with endotoxin or lipopolysaccharide (LPS). As a potent inducer of cytotoxicity, TNFα is best known for its role in autoimmune diseases, such as rheumatoid arthritis (RA). Etanercept, a TNF-Fc fusion protein, and infliximab (Remicade), a chimeric anti-TNFα mAb, were the first TNF-targeted therapeutics for RA. Patients treated with anti-TNFs appear to be at slightly increased risk for lymphoma, lending credence to the idea that TNFα plays a role in suppressing cancer development.

The potent activity of recombinant TNFα against human tumor xenografts led to clinical trials involving systemic administration for solid tumors. It turned out to be highly toxic, causing severe liver toxicity, multiorgan failure, and vascular collapse. Subsequent trials attempted to localize delivery of TNFα to an isolated limb for melanoma; these were also unsuccessful. It is now known that both TNF-R1 and TNF-R2 are abundantly expressed on endothelial cells, suggesting inhibition of angiogenesis as a possible mechanism of efficacy and the likely cause of vascular toxicity (Daniel and Wilson 2008).

Subsequent strategies to deliver TNFα have been based on the fusion of TNFα to an antibody against a tumor antigen, to a ligand that binds a tumor cell-surface receptor, or to pegylated gold nanoparticles that may home to a tumor (Bremer et al. 2009). Examples in phase II include L19-TNFα, a fusion to an anti-oncofetal fibronec-tin ED-B domain antibody, for sarcoma, and NGR-hTNF, a vascular disrupting agent discussed below. The most advanced TNFα delivery system is golnerminogene pradenovec (TNFerade, Ad GV.EGF.TNF.11D), a replication-defective adenoviral vector encoding TNFα with a radiation-sensitive promoter, which has produced encouraging long-term survival in esophageal cancer and is currently in phase III pancreatic cancer trials.

Another approach has been to develop an agonist antibody to activate a receptor. The first choice was FAS (APO-1/CD95), another death domain-containing TNF-family receptor. The FAS/FAS ligand system had been well-studied and was known to be important in the elimination of aged T cells and in cell-killing activity of CTLs. Unfortunately, in preclinical animal models, hepatotoxicity was so severe that no human trials were attempted (Daniel and Wilson 2008).

8.7.2   TRAIL

Despite the early failures, the search within the TNF superfamily for an efficacious but nontoxic therapeutic has continued and the TRAIL system has emerged as the current promising candidate. TRAIL (TNF-related apoptosis-inducing ligand, Apo2L) is again a homotrimeric membrane-bound ligand that also exists in a soluble form. Its normal functions seem to be in immunosurveillance, interferon-mediated innate immune responses, NK activity, and suppression of autoimmunity. Therapeutically, its relevant receptors are DR4 (death-receptor 4, TRAILR1) and DR5 (KILLER, TRAILR2), both of which contain the internal death domain. Binding of TRAIL to these receptors induces cross-linking, DISC formation, and activation of the extrinsic caspase pathway through caspase 8 and 10, resulting in apoptosis of the target cell. The intrinsic caspase pathway appears to play a critical role as well. Of vital importance is the fact that apoptosis is independent of p53, meaning that the 50% of tumors in which p53 is mutated or absent may still be susceptible to a TRAIL therapeutic (Takeda et al. 2007; Wang 2008). Moreover, preclinical results strongly indicate that many tumor cells are particularly susceptible to TRAIL-induced apoptosis, while normal cells are generally quite resistant, giving hope for a wide therapeutic window. The reason is unclear, although it seems reasonable that the process of becoming a tumor cell may be a priming event for apoptosis through DR4 and DR5. Finally, both traditional chemotherapeutics and newer targeted small-molecule therapies appeared to sensitize tumor cells and xeno-graft tumors to TRAIL therapeutics, indicating that these therapeutics may fit well in the clinic in combination with standards-of-care.

The recombinant ligand rhApo2L/TRAIL (dulanermin) has progressed to phase II trials in both hematological and solid malignancies. Following early structural manipulations, the current version has demonstrated relatively low toxicity and mini-mal hepatotoxicity in both preclinical and clinical studies, thereby setting it apart from TNFα and FAS. Clinical responses in NSCLC and NHL were achieved in the phase I trials. Meanwhile, agonist antibodies to DR4 (mapatumumab) and DR5 (apomab, lexatumumab, conatumumab, tigatuzumab) have all reached phase II trials as well. Objective responses with low toxicity in solid and hematological tumors were observed in the phase I trials of all of these antibodies (Ashkenazi 2008).

It is unclear whether any of these therapeutics will turn out to be superior to the others. The recombinant ligand has the advantage of activating both DR4 and DR5. Conversely, agonist antibodies to DR5 have been shown not only to induce apoptosis directly, but also to stimulate Fc-mediated effector cell-killing of tumor cells; furthermore, they also induce tumor-specific effector and memory T cells that may provide immunity to protect against tumor recurrence.

8.7.3 Other TNF-Family Members

Some TNF-family members can also play the opposite role of promoting or maintaining tumor growth, rather than inducing apoptosis. One example is the CD40/CD40-ligand axis, which is active in many B-cell malignancies. Two antagonist antibodies (dacetuzumab and lucatumumab) that target the CD40 ligand have reached phase II trials in hematological malignancies.

Other TNF receptor family members under early investigation as targets of apoptosis-inducing agonist antibodies include the lymphotoxin beta receptor (LTBR) and Fn14, the TWEAK receptor. Meanwhile, an additional route to efficacy has been suggested by experiments in which tumor-bearing mice treated with expression vectors encoding the ligand LIGHT developed long-term antitumor immunity.

8.8 Tumor Microenvironment and Angiogenesis

With the therapeutics discussed so far, the strategy has been to attack the malignant tumor cells. However, it has become evident that the tumor microenvironment (TME) plays an important role in supporting the tumor (Joyce 2005). Nonmalignant cells in the tumor, including endothelial cells (ECs), lymphocytes, fibroblasts, and macrophages, may be recruited by the malignant tumor cells to provide support which is critical for tumor growth and survival. Targeting these nonmalignant components may be of significant benefit. Since most of the interactions are extracellular, the TME is particularly appropriate for protein-based therapies.

8.8.1 Angiogenesis

The most prominent example of a therapy aimed at the TME is bevacizumab, an antibody recognizing vascular endothelial growth factor (VEGF) (Grothey and Galanis 2009). It is believed that a tumor cannot grow beyond a certain size before it must recruit its own supply of nutrients and oxygen. As it reaches the limiting size and becomes hypoxic, it begins to secrete factors such as VEGF that stimulate the growth of blood vessels into the tumor, a process called angiogenesis. Bevacizumab has been approved as an anti-angiogenesis agent in combination with chemotherapy for metastatic colorectal cancer, NSCLC, Her-2 negative breast cancer, and glioblastoma. It has failed in some indications, such as early-stage colon cancer, but other approvals are expected.

As a result of the success of bevacizumab, other therapies are aimed at disrupting the VEGF signaling axis. Ramucirumab is an antibody that recognizes the VEGF receptor KDR (VEGFR2) and blocks binding of VEGF. It is currently in phase III trials for breast cancer and other late-stage trials. Aflibercept is known as a "VEGF trap." It is a recombinant protein consisting of portions of the two VEGF receptors 1 and 2 and acts as a decoy to soak up VEGF. The drug is currently in phase III trials for NSCLC, prostate and colorectal cancer. Small-molecule inhibitors of the intracellular kinase domain of the VEGF receptors are also in development.

Angiogenesis is a complicated process, involving significant redundancy. Hence, a profusion of experimental protein therapeutics is aimed at the myriad other inducers of angiogenesis, including Flt-3 (fms-related tyrosine kinase-3) (IMC-EB10, in phase I), bFGF (basic fibroblast growth factor) and the FGF receptors, and the angiopoietins and their receptor TEK/Tie2. A curious new target is sphingosine-1-phosphate (S1P), a bioactive lipid that signals extracellularly by binding several G-protein-coupled receptors and acts as an extremely potent stimulator of endothelial cell migration (Shida et al. 2008). Sonepcizumab is an anti-S1P antibody that depletes S1P from the blood and is currently in phase I.

While antibodies are popular, novel protein-based entities are also being tested as angiogenesis inhibitors. AMG-386 is a "peptibody" composed of a peptide of Tie2 fused to an Fc domain; it acts as a decoy receptor to soak up the angiopoietin ligands (Herbst et al. 2009). CT-322 is an "adnectin," a small domain of human fibronectin engineered to include a VEGFR2-binding domain that antagonizes the binding of VEGF; its small size promises improved tumor penetration over conventional antibodies (Dineen et al. 2008). These therapeutics have progressed to phase II in solid tumors and in glioblastoma, respectively. Other therapeutics are designed to inhibit facilitators of angiogenesis, including the integrins and matrix metalloproteinases (discussed below). Moreover, the original paradigm of inhibiting blood vessel growth into the tumor is only part of the story. It is now clear that part of the MOA of bevacizumab is to "normalize" a disordered vasculature within the tumor and reduce interstitial fluid pressure, thereby acting as an adjuvant to traditional chemotherapy by enhancing the delivery of the drug to the tumor cells (Duda et al. 2007).

In addition to angiogenesis inducers such as VEGF, against which the aforementioned therapeutics are directed, there are at least 27 known naturally occurring inhibitors of angiogenesis in humans (Nyberg et al. 2005). The presence of inducers and inhibitors led to the hypothesis that a delicate balance exists between formation and suppression of new blood vessels. Some investigators have referred to an angiogenic switch, which ordinarily prevents the nourishment of tumors. Natural inhibitors of angiogenesis fall into two classes: matrix-derived factors that are proteolytically cleaved fragments of extracellular matrix or basement membrane and nonmatrix-derived factors. Endostatin, a fragment from the C-terminal domain of collagen XVIII from blood vessels, and angiostatin, a cryptic fragment of plasminogen, are notable examples of these two classes; their discovery in part validated the theory of angiogenesis in oncology. It was hoped that recombinant endostatin and angiostatin would turn out to be useful therapies. However, in more than a dozen years, both have only progressed to phase II trials. Meanwhile, almost no therapeutic progress has occurred with any of the other natural inhibitors.

8.8.2 VDAs

A new class of therapeutics known as vascular disrupting agents (VDAs) are distinguished from angiogenesis inhibitors by their ability to cause a catastrophic collapse of the tumor vasculature within minutes to hours of drug administration (Hinnen and Eskens 2007). So far, most of these agents are small molecules, especially the microtubule-depolymerizing combretastatins, but analogous antibodies and endothelial cell-targeted fusion proteins are also under investigation. Novel fusion proteins in phase II for highly vascularized tumors include: NGR-hTNF (Gregorc et al. 2009), which directs apoptosis-inducing $TNF\alpha$ to endothelial cells by means of a homing peptide that binds aminopeptidase N/CD13, and L19-IL2 (Wagner et al. 2008), which directs IL2 to ECs through a single-chain Fv recognizing extra-domain B of fibronectin. Also in phase II is bavituximab, a mAb targeting phosphatidylserine on the surface of ECs (He et al. 2007).

8.8.3 TAMS

Tumor-associated macrophages (TAMs) are another cell target in the TME (Sica et al. 2008). Leukocytes can make up 50% of the tumor mass and TAMS are a major component. It has been shown that monocytes migrate into the tumor in response to chemokines, differentiate in response to M-CSF, and become polarized to an M2-phenotype in response to $TGF\beta$, IL-10, and other tumor-derived factors. These tumor-supported M2-macrophages, frequently residing in hypoxic areas of the tumor, have an immunosuppressive, proangiogenic and prometastatic phenotype, due to their secretion of growth factors, immunosuppressive cytokines, and matrix metalloproteinases. High numbers of TAMs are associated with poor prognosis of human cancers and animal models have demonstrated that reducing TAM populations reduces both the size of tumors and the number of metastases. Research is still at an early stage: in preclinical development are antibodies to M-CSF and macrophage migration inhibitory factor (MIF, a protumorigenic stimulator of macrophages and other leukocytes) and $TGF\beta$ (see below), as well as small-molecule kinase inhibitors of the M-CSF receptor c-FMS. In addition, a combination of an anti-IL-10 receptor antibody with the TLR9 ligand CpG has been shown to reprogram M2 macrophages to an M1 immunostimulatory Th1-oriented phenotype, relieving the paralysis of tumor-infiltrating dendritic cells and causing rapid tumor necrosis (Guiducci et al. 2005). Clinical anti-$TGF\beta$ and anti-MMP programs are discussed below.

8.8.4   CAFs

Carcinoma-associated fibroblasts (CAFs) or myofibroblasts, which display smooth muscle differentiation markers (Bhowmick et al. 2004), also appear to support tumor growth, particularly through the secretion of various protumorigenic growth factors. An early program with sibrotuzumab, an antibody to fibroblast activation protein, appears to have languished, but antibodies to stromal cell-derived factor-1 (SDF-1, CXCL12, the CXCR4 receptor) and TGFβ may prove useful against CAFs.

One general advantage of targeting stromal cells is that they are (theoretically) normal and genetically stable, and therefore will not acquire mutations that circumvent the therapeutic.

8.9   Metastasis

Metastasis refers to the spread of tumor cells to sites distant from the original tumor. In most fatal solid cancers, proliferation of metastases is the ultimate cause of death. Hence, it would seem that therapies directed at preventing metastasis should be of paramount importance in research. The conundrum is that early clinical trials almost always involve patients with widespread metastatic disease, thereby obviating any road to success with an agent intended to prevent metastasis. When trials can be designed with patients who show no signs of metastasis, those trials are likely to require a lengthy period of time until a definitive endpoint can be reached. Moreover, there is continued debate about whether metastasis is a late phenomenon or whether micrometastases lie dormant even in early-stage patients. Despite the difficulties, research is ongoing. Metastasis occurs when epithelial cells undergo an epithelial-to-mesenchymal transition (EMT), lose their attachment to neighboring cells and to the basement membrane, degrade the matrix, and travel to distant sites through the vasculature or lymphatic system (Yang and Weinberg 2008). Growth and migratory factors, cell adhesion proteins, and proteases are some of the extracellular players which may prove to be good targets for therapeutic proteins such as antibodies (Iiizumi et al. 2008). The HGF (scatter factor)/c-Met pathway (discussed above) and the TGFβ pathway (discussed below) are two powerful inducers of metastasis, while several of the integrins (discussed below) are critically involved in the loss of cell–cell contact and recolonization in distant organs. A number of natural suppressors of metastasis have also been identified, but these are generally intracellular molecules.

8.9.1   RANKL and Denosumab

A promising new metastasis-related therapy is denosumab (Burkiewicz et al. 2009), an antibody directed against the RANK (receptor activator of NFkB) ligand (RANKL).

Binding of RANKL to RANK is required for the activation and maintenance of osteoclasts, the cells primarily responsible for bone resorption. Denosumab blocks the binding of RANKL, thereby preventing osteoclast development and bone loss. An obvious therapy for osteoporosis, denosumab has also shown clinical promise for two applications in oncology. In phase III trials of patients receiving androgen-deprivation therapy for nonmetastatic prostate cancer, denosumab significantly increased bone mineral density in the lumbar spine and at nonvertebral sites and achieved a 50% reduction in vertebral fractures compared to placebo. Similarly, in nonmetastatic breast cancer patients receiving estrogen reduction therapy, bone mineral density was increased at multiple sites relative to placebo. Meanwhile, multiple phase III trials are ongoing in prostate, breast, and other cancers to show whether denosumab can prevent the occurrence of bone metastases and/or suppress the bone destruction associated with bone metastases. Early trials for direct treatment of the rare giant cell tumor of the bone are also in progress.

8.9.2 MMPs

In order to become invasive, a cell must cut its way through a basement membrane or the surrounding stroma. The matrix metalloproteinases (MMPs) are a family of at least 28 zinc- and calcium-dependent endopeptidases that are involved in extracellular matrix degradation, tissue remodeling, and growth factor/cytokine activation. A wealth of data from cell culture, transgenic animals, and MMP-knockout mice has shown the importance of various MMPs in cancer, particularly in tumor invasion, metastasis, and neovascularization. Patient data have also linked overexpression of some MMPs to poor prognosis. For nearly two decades, the MMPs have been very promising targets in oncology; yet, clinical results with a plethora of small-molecule and peptide-derived MMP inhibitors have been dismal (Fingleton 2008). Redundancy of MMP function appears to be one cause of failure. But research continues and the first anti-MMP antibody has entered the fray. DX-2400 is a selective antagonist of MMP-14, an important enzyme in collagen degradation. Still in the discovery stage, the antibody has proven to inhibit angiogenesis and slow tumor progression and metastasis formation in animal models (Devy et al. 2009). Interestingly, MMP-14-null mice are the only MMP knockouts to show a severe stand-alone phenotype, emphasizing the need for caution; nonetheless, clinical success with this MMP antibody will open the door to many others.

8.9.3 TGFβ

Transforming growth factor-β (TGFβ) illustrates the complexity of any approach to altering cancer signaling pathways (Pennison and Pasche 2007). Three independent TGFβ isoforms exist. They bind the TGFβ receptor, a tetramer of two type I, and two type II transmembrane serine-threonine kinases. Positive and negative signaling occurs through a family of transcription factors called Smads, including Smad4 (DPC4), a tumor suppressor that is frequently deleted in pancreatic and colorectal cancer and familial juvenile polyposis. TGFβ is involved in many processes, including proliferation, differentiation, development, wound healing, and extracellular matrix synthesis. In cancer, it is a major inducer of EMT and metastasis. The growth of normal epithelial and hematopoietic cells and early-stage carcinomas is generally inhibited by TGFβ. However, this inhibition is often lost in midstage tumors, while in late-stage tumors TGFβ may actually promote invasion and metastasis (Tang et al. 2003). Some aggressive tumors secrete TGFβ themselves, resulting in autocrine stimulation of growth of tumor cells that have undergone an EMT, while also suppressing effector cells involved in an antitumor immune response. It is anticipated that anti-TGFβ therapeutics will therefore be effective in late-stage cancers. Nonetheless, preclinical results indicate that even within a single tissue type, some tumors may be susceptible to a TGFβ inhibitor, while the growth of others may be accelerated, thereby highlighting the importance and the challenge of patient selection in clinical trials. While much preclinical work has been done with a soluble decoy version of the receptor TGFβRII to scavenge TGFβ, the clinical therapeutics include neutralizing antibodies to all isoforms of TGFβ, as well as antisense oligonucleotides and small-molecule inhibitors of the TGFβRI kinase. The anti-TGFβ antibody GC-1008 achieved stable disease or better in 5 of 21 advanced malignant melanoma patients in phase I and is now in phase II for melanoma and renal cell carcinoma.

8.10   Integrins

The integrins are a large family of transmembrane heterodimers that mediate cell adhesion, both to other cells and to proteins in the extracellular matrix, such as laminin, collagen, and fibronectin. Eighteen distinct α subunits and eight β subunits can combine to form 24 αβ integrins that connect the internal cytoskeleton to the extracellular space and generate signals in both directions across the plasma membrane. Five integrin antagonists are currently on the market for nononcology indications, thus validating the integrins as disease targets. In cancer, integrins on tumor cells play a role in invasion and metastasis, processes which involve the loss or reorganization of proper cell adhesion. Integrins on endothelial cells facilitate tumor-induced angiogenesis, while a different set of integrins regulate lymphangiogenesis, the analogous formation of new lymphatic vessels into a tumor (Garmy and Varner 2008). Some integrins on both normal and tumor cells can be dependence receptors; they trigger apoptosis upon loss of attachment to their ligand. Integrins have garnered much interest as oncology targets and various antibodies are in development to inhibit specific functions, depending on the integrin (Tucker 2006). The most prominent examples include:

(1) integrins αvβ3 and αvβ5 (targeted by intetumumab) or integrin αvβ3 alone (targeted by etaracizumab, possibly on hold), for solid tumors to inhibit angiogenesis,

invasion, and metastasis; (2) integrin α5β1 (targeted by volociximab and PF-04605412) for solid tumors to inhibit angiogenesis; (3) integrin α4β1, a leukocyte adhesion molecule for multiple myeloma (targeted by natalizumab); and (4) integrin αvβ6. Antibodies to osteopontin, which binds the leukocyte integrins α4β1, α9β1, and α9β4 and is frequently overexpressed in tumors, are in discovery; these antibodies may also inhibit lymphangiogenesis. A number of integrins recognize a surprisingly small domain on their ligands, termed the RGD motif (arginine, glycine, aspartic acid); hence, small peptides and small molecules are also being developed to neutralize this interaction.

On the subject of cell adhesion, other current antibody targets include EpCAM (targeted by catumaxomab, discussed below; adecatumumab, in phase II for solid tumors; edrecolomab, now withdrawn after several trials; and several toxin-conjugated antibodies), VE-cadherin, and P-cadherin.

8.11 Cancer Stem Cells

Cancer stem cells. One theory of cancer suggests that a minor population of cells retains the capacity for limitless self-renewal while also giving rise to the majority of cells in the tumor, which have limited growth potential. This minor population consists of the cancer stem cells (CSCs). In theory, the CSCs are genetically and phenotypically different from the bulk of the tumor. The CSCs may be derived from normal tissue stem cells that have been oncogenically transformed, as for example, the stem cells of the colon crypts, or they may be cells which have been selected to have "stem cell-like" characteristics during the process of oncogenesis. In either case, the CSCs are also thought to respond differently or not at all to standard chemotherapies and radiotherapy. Thus, when the bulk of a tumor is killed by initial therapy, the CSCs survive to reestablish the tumor or the distant metastases.

Proponents of the hypothesis suggest that CSC-specific therapeutics are needed in order to elicit real cancer cures. Modeled on the elucidation of normal hematopoietic stem cells, the CSC field has been largely defined by antibodies that recognize cell-surface antigens by fluorescence-activated cell sorting (FACS). For example, in AML, a fairly large percentage of patient cells are CD34+CD38+ and can form colonies in agar, but only the very rare CD34+CD38- cells can initiate leukemias in mice (Dick 2008). The first step is to identify specific cell-surface markers that are found on the stem cells for each tumor type, but not on the normal cells or on the bulk tumor cells. Putative stem-cell marker combinations have been found for several tumors and research is ongoing to hone the identification with antibodies to additional markers (Klonisch et al. 2008). The second step will be to develop therapeutics, and a number of groups have begun the process of finding anti-CSC monoclonal antibodies based on the specific surface antigens or on extracellular signaling pathways involved in self-renewal (e.g., Wnt, Hedgehog, Notch).

8.12 All the Rest

The range of different protein therapeutic modalities extends beyond the scope of this review and seems to be limited only by investigators' imaginations. A few unusual approaches are briefly noted here.

8.12.1 Enzymes

L-asparaginase is an enzyme that is used as an important therapy in acute lymphoblastic leukemia (ALL). Lymphoblast leukemia cells cannot make the nonessential amino acid asparagine and must take it from the circulation. Asparaginase degrades asparagine, thus depriving the leukemia cells, leading to inhibition of their protein synthesis. Pegylated asparaginase is now in common use to prevent allergic reactions and research continues on modifications to extend circulating half-life and further improve pharmacokinetics (Masetti and Pession 2009).

Ranpirnase is an RNase that binds to tumor cell-surface receptors and is taken into cells by energy-dependent endocytosis, where it degrades RNA, particularly tRNA, thus inhibiting protein synthesis and triggering the intrinsic apoptotic pathway. It has anti-tumor activity in vitro and in animal models. Ranpirnase was carried through phase III trials in unresectable malignant mesothelioma. Although it failed to show a statistically significant overall survival benefit, it did show a statistical benefit in patients who had previously failed one chemotherapy. A confirmatory phase III trial is planned for that patient subpopulation. Meanwhile, second-generation therapeutics are being designed in which the RNase is linked as an immunoconjugate to internalizing antibodies (Krauss et al. 2008).

8.12.2 Peptides

Various peptide structures are under investigation. One interesting variation is the "stapled peptide." Stapled peptides are synthesized with a chemical cross-link to lock them into their functional alpha-helical shape. They are protease-resistant in the serum and maintain their cell permeability, allowing them to be used against intracellular targets. BH3-domain stapled peptides have been employed to trigger apoptosis in cancer cells by activating the proapoptotic bcl-2 family member BAX (Walensky et al. 2004).

8.12.3 Aptamers

To conclude the discussion of protein therapeutics on a contrary note, it may be useful to mention another type of macromolecule. An aptamer is not a protein at all, but rather an oligonucleotide that folds into a specific, stable 3-dimensional conformation, making it capable of binding another macromolecule with high affinity and specificity. It thus can function like an antibody (Barbas and White 2009). With the approval of pegaptanib (Macugen), an aptamer that binds VEGF, for age-related macular degeneration in 2004, aptamers gained therapeutic legitimacy. AS-1411, an aptamer against nucleolin, is currently in clinical trials for acute myelogenous leukemia and renal cell carcinoma.

8.13   Antibody Engineering

Complementing the search for new molecular targets in oncology, much work has gone into advancing the selection and structural design of protein therapeutics. For example, the site-directed attachment of polyethylene glycol polymers to a protein, known as pegylation, can provide a longer half-life in vivo, allowing for less frequent dosing. With particular regard to antibodies, investigators have defined the function of every part of the antibody structure and can now alter, in a rational way, the specificity, immunogenicity, mechanism-of-action, and pharmacokinetics of the potential therapeutic (Liu et al. 2008; Presta 2008). Some of that progress is summarized here. The structure of a normal antibody with its binding domains is depicted in Fig. 8.1a.

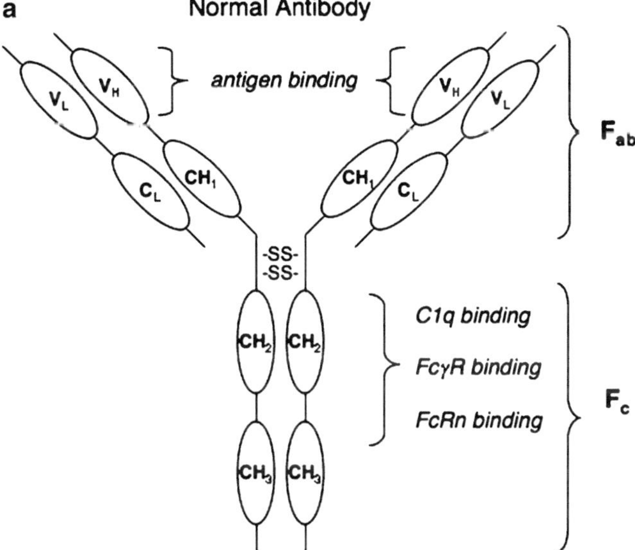

Fig. 8.1 Antibody formats. (**a**) A normal bivalent antibody, with sites of interaction. Only the hinge-region disulfides are depicted, (**b**) monovalent single-chain Fv, (**c**) bivalent bispecific quadroma antibody, (**d**) bispecific tandem scFv, (**e**) two-chain bispecific diabody, and (**f**) tetravalent bispecific antibody

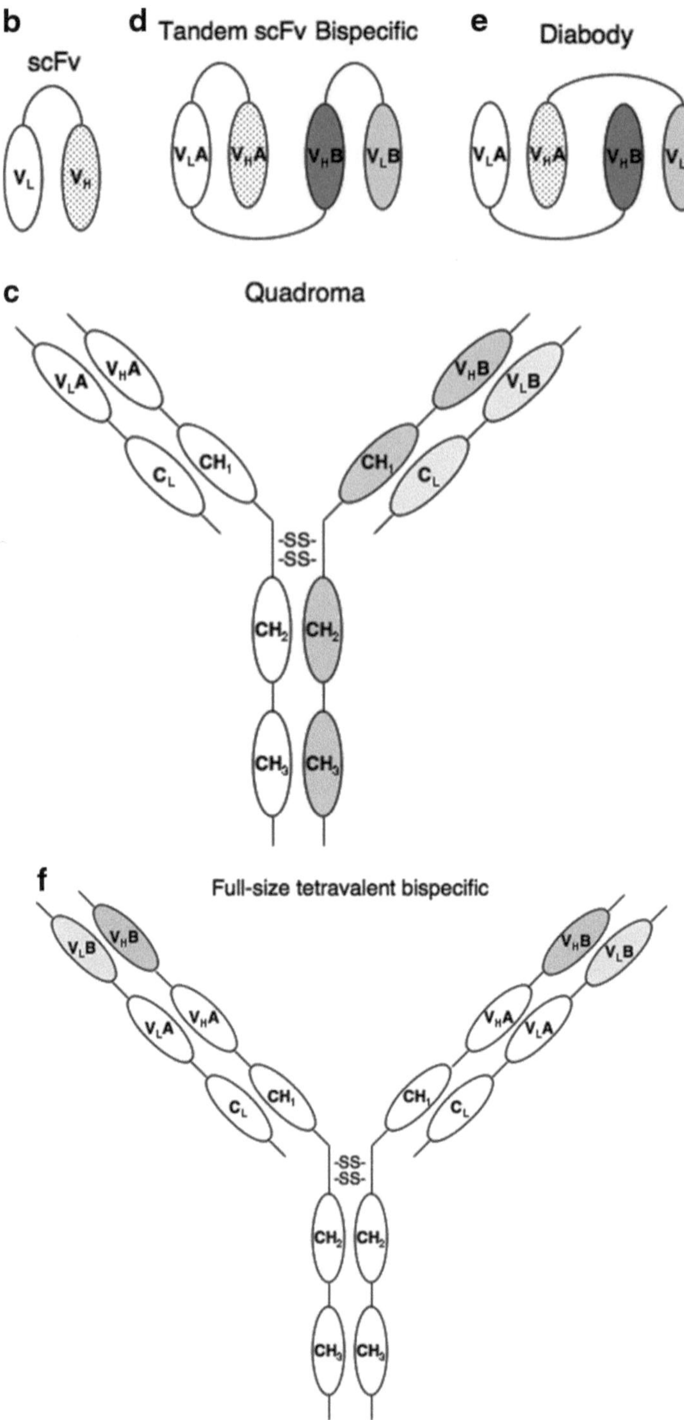

Fig. 8.1 (continued)

8.13.1 Antigen Binding

The first goal is usually to generate an antibody with extremely high affinity and specificity for the target, in order to maximize efficacy while limiting off-target toxicity. In some cases, binding of a promising antibody can be improved after selection by direct substitution of residues in the complementarity determining regions (CDRs). In practice, increasing the diversity of the antibody pool at the outset provides a better opportunity to find a useful therapeutic. Traditional generation of monoclonal antibodies by immunization of rodents is still a mainstay. However, a truly universal panel of antibodies is never achieved, because of interference by the host biology, in which some antigens or epitopes are seen as self. This is especially true for extracellular antigens and for antigens where the homology between rodents and humans is high.

One solution has been to create phage-, yeast-, and ribosome-display libraries. In a phage-display library, unselected human V_H and V_L sequences are inserted into a phage genome and expressed on the phage surface. The library is "panned" with the antigen of choice. Those phage that bind are isolated and the DNA sequence is cloned out. Using Fab fragments generated in this manner, a complete antibody can be constructed. Antibody selection is therefore free from the constraints of *in vivo* generation. Adalimumab, an anti-TNF antibody for rheumatoid arthritis, was the first approved antibody generated by phage display. Efforts to increase the sequence diversity have resulted in libraries containing tens of billions of individual antibodies, from which the selection can occur (Mondon et al. 2008).

8.13.2 Humanization

The initial attempts at making antibody therapeutics utilized mouse monoclonals, derived in the traditional fashion by immunizing a mouse with the antigen of choice and fusing the spleen to a myeloma cell line. It became apparent that these antibodies were immunogenic when transferred to another species, such as the human patient. The immune response against the therapeutic antibody led to rapid elimination and occasionally dangerous inflammatory reactions. The first solution was primatization: the replacement of certain parts of the constant regions with analogous fragments from nonhuman primates. Rituximab (see above) is a primate/murine chimera. The second solution was partial humanization, accomplished by combining mouse variable domains with human constant domains. Complete humanization involved grafting only the rodent CDRs of a promising antibody into a human framework. Alemtuzumab (see above) was generated in this fashion. Nowadays, humanization is a more complex process involving a careful determination of the binding contribution of each amino acid both in and outside of the CDRs, followed by several cycles of generation of variants until maximum binding is achieved with a minimum number of nonhuman residues (Almagro and Fransson 2008). Alternatively, several empirical approaches can now generate fully human antibodies at the start. These include libraries based on human antibody sequences, such as the phage-display technique described above, and traditional antibody generation in a transgenic mouse host in which the mouse immunoglobulin loci have been replaced with human germline immunoglobulin loci. While some antibodies in late-stage oncology clinical trials are still chimeras, it is likely that in the future only fully human or humanized antibodies will be utilized as therapeutics.

8.13.3   Fc Engineering

While the variable domains of an antibody are important for antigen recognition, the constant regions determine how the immune system interacts with the antibody. Hence, a great deal of engineering has been directed at the Fc portion of the antibody, to improve both effector function and the pharmacokinetics in circulation. For ADCC, the Fc domain of the four human IgG isotypes interacts with a set of receptors known as FcγR, found on effector cells such as NK cells and macrophages. Both activating and inhibitory FcγRs exist. Certain amino acid substitutions within the Fc domain, as well as certain alterations in glycosylation, result in a higher affinity for activating FcγR and a lower affinity for inhibitory FcγR. When engineered into established antibodies, some of these mutations significantly enhance ADCC. Likewise, the Fc domain interacts with the C1q complex to initiate the classical complement cascade, ultimately resulting in the formation of the membrane attack complex that kills the target cell. Hence, other mutations in the Fc domain can improve CDC. The efficacy of either rituximab or trastuzumab is dramatically reduced in tumor xenograft experiments carried out in mice deficient for either FcγR or complement, demonstrating the importance of these two pathways. Therefore, for those therapeutic antibodies where ADCC and/or CDC are likely to be important or desirable, the Fc mutations can be introduced; whether they will actually improve efficacy in the clinic has yet to be determined. However, retrospective analysis of patient samples does suggest that FcγR polymorphisms that affect Fc binding do have different outcomes in response to ADCC-generating therapeutics such as trastuzumab (Musolino et al. 2008).

8.13.4   Effectorless Antibodies

In other cases, the opposite approach has been taken with regard to Fc function. For therapeutic antibodies in which the primary MOA is disruption of ligand binding or induction of apoptosis rather than ADCC or CDC, it has been shown that effector function may even be undesirable, due to off-target inflammation and normal organ toxicity. Solutions include amino acid substitution or deglycosylation of the Fc domain to eliminate binding to FcγR; utilization of an IgG4 subtype antibody, because IgG4 binds very poorly to both C1q and the important FcγRs; and employment of antibody fragments such as Fab or scFv (see below) that lack an Fc domain entirely.

8.13.5   Pharmacokinetics

Another goal of antibody engineering is to be able to select the desired pharmacokinetic profile by manipulation of the half-life in circulation. The Fc portion of the IgG immunoglobulin family also has a binding site for FcRn, the neonatal Fc receptor family, which is found on many cell types and which earlier in life is responsible for transporting IgG from mother to fetus across the placenta. Binding of IgG to FcRn results in internalization and catabolism; hence, this complex essentially regulates serum IgG concentrations. The FcRn binding site involves residues located at the interface between the CH2 and CH3 domains of the Fc region of the antibody, distinct from the FcγR and C1q binding sites. Mutation of certain of these residues results in increased binding, decreased serum concentration, and reduced half-life. This more rapid elimination is useful for immunoconjugates, diagnostic and screening antibodies, and other therapeutics where off-target toxicity is problematic. Other approaches to reducing half-life include truncation or elimination of the Fc domain or utilization of antibody fragments (Fab, scFv, etc., see below), while increased half-life may be achieved by conjugation of albumin or alteration in glycosylation.

8.14   New Antibody Structures and Bispecific Antibodies

8.14.1   One-Armed Antibodies and scFv

An ordinary antibody consists of two identical heavy and light chain pairs, creating two identical recognition sites for a single antigen (Fig. 8.1a). Cell fusion and recombinant DNA technology have facilitated the development of engineered antibodies that do not exist naturally. An example is the "one-armed" or monovalent antibody, which possesses only a single heavy-light chain pair. A two-armed divalent antibody can bind two antigens on the cell surface. With some transmembrane receptors, such binding results in dimerization of the receptors, thus mimicking the activating action of ligand binding. Since the antibodies are efficacious, it is clear that interruption of ligand binding wins out over the counteraction of this "agonism." Nonetheless, intuition suggests that efficacy would be improved if the agonism were eliminated. The c-Met antibody MetMab (described above) is a one-armed antibody.

Further reduction in coding sequence yields a single-chain Fv (scFv) antibody, which consists only of the antigen-binding variable region of a heavy chain and the corresponding light chain, joined by a linker region to form a single polypeptide (Fig. 8.1b). The small size of scFvs allows better tumor penetration and a shorter half-life, which can be advantageous for immunoconjugates and imaging antibodies. The single polypeptide format facilitates selection from phage-display libraries.

8.14.2 Bispecific Antibodies

Normal antibodies and these smaller versions recognize a single antigen. In contrast, a bispecific antibody recognizes two separate antigens (Chames and Baty 2009). There are many formats. Full-size bispecifics, called quadromas, can be formed by the pairing of two distinct heavy-light chain pairs after fusion of two hybridomas (Fig. 8.1c). Smaller bispecifics lack some or all of the antibody constant regions. One format is the "tandem scFv," in which two distinct single-chain variable regions have been linked by a flexible peptide (Fig. 8.1d). A second format is the bispecific diabody, in which two mixed variable-domain polypeptides (i.e., V_HA-V_LB and V_HB-V_LA) are expressed in the same cell and associate to create two antigen-binding sites (Fig. 8.1e). Larger bispecific minibodies include the C_H3 and antibody hinge regions to serve as oligomerization domains for two distinct variable domains. Full-size tetravalent bispecifics, encoding two variable recognition domains for each antigen with the second set of variable domains appended to the N-termini of the first set, have also been made (Fig. 8.1f).

Many therapeutic strategies underlie the use of bispecific antibodies (Kufer et al. 2004; Chames and Baty 2009). The original strategy was to induce a polyclonal T-cell response against the tumor by joining a recognition site for a tumor antigen to a recognition site for the T-cell receptor CD3. In this manner, cytotoxic T cells, which do not express the Fcγ receptor, could be drawn into close proximity with tumor cells through the bispecific antibody bridge. Blinatumomab is one such single-chain bispecific antibody, recognizing CD3 and the B-cell tumor antigen CD19. Currently in Phase II for NHL, it has shown impressive tumor regressions and clearance of tumor cells from patient liver and bone marrow. Catumaxomab (Shen and Zhu 2008) is a "trifunctional" bispecific antibody: it has recognition domains for the epithelial cell adhesion molecule EpCAM and CD3 and also retains its Fc domain. In addition to T cells, it recruits NK and macrophage effector cells to the tumor. The NK cells and macrophages act directly on the tumor through ADCC and also generate a costimulatory signal to enhance the T-cell response to the tumor. Catumaxomab has been approved in Europe for malignant ascites in EpCAM-positive carcinomas and is in multiple U.S. trials. A notable characteristic of these and similar bispecific therapeutics is that clinical results have been achieved at doses far lower than the doses used for traditional therapeutic antibodies.

More recent bispecific approaches involve single antibodies recognizing two antigens on the same tumor (e.g., EGFR and IGF-1R) (Lu et al. 2005) or even two antigens in a tumor-cell-surface dimer (e.g., erbB2/erbB3) (Robinson et al. 2008). The hypothesis is that using a single bispecific in place of two monospecific antibodies may prove synergistic in tumor cell-killing and also achieve certain economies in terms of drug administration. Alternatively, a bispecific may recognize two epitopes on the same antigen, with the intention of increasing avidity while reducing the incidence of resistance developing through mutation of one epitope. Going one step further, a variant of trastuzumab has been developed in which a single antigen-binding site recognizes VEGF as well as HER2 (Bostrom et al. 2009).

References

Almagro JC, Fransson J (2008) Humanization of antibodies. Front Biosci 13:1619–1633

Ashkenazi A (2008) Directing cancer cells to self-destruct with pro-apoptotic receptor agonists. Nat Rev Drug Discov 7:1001–1012

Atkins MB (2002) Interleukin-2: clinical applications. Semin Oncol 29(3 Suppl 7):12–17

Barbas AS, White RR (2009) The development and testing of aptamers for cancer. Curr Opin Investig Drugs 10:572–578

Baselga J, Swain SM (2009) Novel anticancer targets: revisiting ERBB2 and discovering ERBB3. Nat Rev Cancer 9:463–475

Begley J, Ribas A (2008) Targeted therapies to improve tumor immunotherapy. Clin Cancer Res 14:4385–4391

Bhowmick NA, Neilson EG, Moses HL (2004) Stromal fibroblasts in cancer initiation and progression. Nature 432:332–337

Borden EC (2005) Review: Milstein Award lecture: interferons and cancer: where from here? J Interferon Cytokine Res 25:511–527

Borden EC, Sen GC, Uze G et al (2007) Interferons at age 50: past, current and future impact on biomedicine. Nat Rev Drug Discov 6:975–990

Bostrom J, Yu SF, Kan D et al (2009) Variants of the antibody herceptin that interact with HER2 and VEGF at the antigen binding site. Science 323:1610–1614

Bremer E, de Bruyn M, Wajant H et al (2009) Targeted cancer immunotherapy using ligands of the tumor necrosis factor super-family. Curr Drug Targets 10:94–103

Burkiewicz JS, Scarpace SL, Bruce SP (2009) Denosumab in osteoporosis and oncology. Ann Pharmacother 43:1445–1455

Castillo J, Winer E, Quesenberry P (2008) Newer monoclonal antibodies for hematological malignancies. Exp Hematol 36:755–768

Chames P, Baty D (2009) Bispecific antibodies for cancer therapy. Curr Opin Drug Discov Devel 12:276–283

Ciardiello F, Tortora G (2008) EGFR antagonists in cancer treatment. N Engl J Med 358:1160–1174

Copier J, Dalgleish AG, Britten CM et al (2009) Improving the efficacy of cancer immunotherapy. Eur J Cancer 45:1424–1431

Daniel D, Wilson NS (2008) Tumor necrosis factor: renaissance as a cancer therapeutic? Curr Cancer Drug Targets 8:124–131

Devy L, Huang L, Naa L et al (2009) Selective inhibition of matrix metalloproteinase-14 blocks tumor growth, invasion, and angiogenesis. Cancer Res 69:1517–1526

Dick JE (2008) Stem cell concepts renew cancer research. Blood 112:4793–4807

Dineen SP, Sullivan LA, Beck AW et al (2008) The Adnectin CT-322 is a novel VEGF receptor 2 inhibitor that decreases tumor burden in an orthotopic mouse model of pancreatic cancer. BMC Cancer 8:352

Dranoff G (2002) GM-CSF-based cancer vaccines. Immunol Rev 188:147–154

Duda DG, Batchelor TT, Willett CG et al (2007) VEGF-targeted cancer therapy strategies: current progress, hurdles and future prospects. Trends Mol Med 13:223–230

Fanale MA, Younes A (2007) Monoclonal antibodies in the treatment of non-Hodgkin's lymphoma. Drugs 67:333–350

Fingleton B (2008) MMPs as therapeutic targets–still a viable option? Semin Cell Dev Biol 19:61–68

Garmy SB, Varner JA (2008) Roles of integrins in tumor angiogenesis and lymphangiogenesis. Lymphat Res Biol 6:155–163

Goldstein D, Laszlo J (1988) The role of interferon in cancer therapy: a current perspective. CA Cancer J Clin 38:258–277

Gregorc V, Santoro A, Bennicelli E et al (2009) Phase Ib study of NGR-hTNF, a selective vascular targeting agent, administered at low doses in combination with doxorubicin to patients with advanced solid tumours. Br J Cancer 101:219–224

Grothey A, Galanis E (2009) Targeting angiogenesis: progress with anti-VEGF treatment with large molecules. Nat Rev Clin Oncol 6:507–518

Guiducci C, Vicari AP, Sangaletti S et al (2005) Redirecting in vivo elicited tumor infiltrating macrophages and dendritic cells towards tumor rejection. Cancer Res 65:3437–3446

He J, Luster TA, Thorpe PE (2007) Radiation-enhanced vascular targeting of human lung cancers in mice with a monoclonal antibody that binds anionic phospholipids. Clin Cancer Res 13:5211–5218

Herbst RS, Hong D, Chap L et al (2009) Safety, pharmacokinetics, and antitumor activity of AMG 386, a selective angiopoietin inhibitor, in adult patients with advanced solid tumors. J Clin Oncol 27:3557–3565

Hinnen P, Eskens FA (2007) Vascular disrupting agents in clinical development. Br J Cancer 96:1159–1165

Hudis CA (2007) Trastuzumab–mechanism of action and use in clinical practice. N Engl J Med 357:39–51

Hynes NE, Lane HA (2005) ERBB receptors and cancer: the complexity of targeted inhibitors. Nat Rev Cancer 5:341–354

Iiizumi M, Liu W, Pai SK et al (2008) Drug development against metastasis-related genes and their pathways: a rationale for cancer therapy. Biochim Biophys Acta 1786:87–104

In addition to these references, the author also made use of the Investigational Drug Database, a product of Thomson Reuters (http://www.thomsonreuters.com), and the NIH clinical trials database (http://www.clinicaltrials.gov)

Jin H, Yang R, Zheng Z et al (2008) MetMAb, the one-armed 5D5 anti-c-Met antibody, inhibits orthotopic pancreatic tumor growth and improves survival. Cancer Res 68:4360–4368

Joyce JA (2005) Therapeutic targeting of the tumor microenvironment. Cancer Cell 7:513–520

Khawli LA, Hu P, Epstein AL (2008) Cytokine, chemokine, and co-stimulatory fusion proteins for the immunotherapy of solid tumors. Handb Exp Pharmacol 2008:291–328

Klonisch T, Wiechec E, Hombach KS et al (2008) Cancer stem cell markers in common cancers - therapeutic implications. Trends Mol Med 14:450–460

Krauss J, Arndt MA, Dubel S et al (2008) Antibody-targeted RNase fusion proteins (immunoR-Nases) for cancer therapy. Curr Pharm Biotechnol 9:231–234

Kufer P, Lutterbuse R, Baeuerle PA (2004) A revival of bispecific antibodies. Trends Biotechnol 22:238–244

Leonard JP, Goldenberg DM (2007) Preclinical and clinical evaluation of epratuzumab (anti-CD22 IgG) in B-cell malignancies. Oncogene 26:3704–3713

Leonard JP, Friedberg JW, Younes A et al (2007) A phase I/II study of galiximab (an anti-CD80 monoclonal antibody) in combination with rituximab for relapsed or refractory, follicular lymphoma. Ann Oncol 18:1216–1223

Lewis Phillips G, Li G, Dugger DL et al (2008) Targeting HER2-positive breast cancer with trastuzumab-DM1, an antibody-cytotoxic drug conjugate. Cancer Res 68:9280–9290

Liu XY, Pop LM, Vitetta ES (2008) Engineering therapeutic monoclonal antibodies. Immunol Rev 222:9–27

Lu D, Zhang H, Koo H et al (2005) A fully human recombinant IgG-like bispecific antibody to both the epidermal growth factor receptor and the insulin-like growth factor receptor for enhanced antitumor activity. J Biol Chem 280:19665–19672

Margolin K (2008) Cytokine therapy in cancer. Expert Opin Biol Ther 8:1495–1505

Masetti R, Pession A (2009) First-line treatment of acute lymphoblastic leukemia with pegaspara-ginase. Biologics 3:359–368

Michieli P, Mazzone M, Basilico C et al (2004) Targeting the tumor and its microenvironment by a dual-function decoy Met receptor. Cancer Cell 6:61–73

Mondon P, Dubreuil O, Bouayadi K et al (2008) Human antibody libraries: a race to engineer and explore a larger diversity. Front Biosci 13:1117–1129

Mould DR, Baumann A, Kuhlmann J et al (2007) Population pharmacokinetics-pharmacodynam-ics of alemtuzumab (Campath) in patients with chronic lymphocytic leukaemia and its link to treatment response. Br J Clin Pharmacol 64:278–291

Musolino A, Naldi N, Bortesi B et al (2008) Immunoglobulin G fragment C receptor polymorphisms and clinical efficacy of trastuzumab-based therapy in patients with HER-2/neu-positive metastatic breast cancer. J Clin Oncol 26:1789–1796

Nyberg P, Xie L, Kalluri R (2005) Endogenous inhibitors of angiogenesis. Cancer Res 65:3967–3979

Pawluczkowycz AW, Beurskens FJ, Beum PV et al (2009) Binding of submaximal C1q promotes complement-dependent cytotoxicity (CDC) of B cells opsonized with anti-CD20 mAbs ofatumumab (OFA) or rituximab (RTX): considerably higher levels of CDC are induced by OFA than by RTX. J Immunol 83:749–758

Pennison M, Pasche B (2007) Targeting transforming growth factor-beta signaling. Curr Opin Oncol 19:579–585

Peruzzi B, Bottaro DP (2006) Targeting the c-Met signaling pathway in cancer. Clin Cancer Res 12:3657–3660

Petrelli A, Circosta P, Granziero L et al (2006) Ab-induced ectodomain shedding mediates hepatocyte growth factor receptor down-regulation and hampers biological activity. Proc Natl Acad Sci USA 103:5090–5095

Pollak M (2008) Insulin and insulin-like growth factor signalling in neoplasia. Nat Rev Cancer 8:915–928

Presta LG (2008) Molecular engineering and design of therapeutic antibodies. Curr Opin Immunol 20:460–470

Ricart AD, Tolcher AW (2007) Technology insight: cytotoxic drug immunoconjugates for cancer therapy. Nat Clin Pract Oncol 4:245–255

Robinson MK, Hodge KM, Horak E et al (2008) Targeting ErbB2 and ErbB3 with a bispecific single-chain Fv enhances targeting selectivity and induces a therapeutic effect in vitro. Br J Cancer 99:1415–1425

Rodon J, DeSantos V, Ferry RJ et al (2008) Early drug development of inhibitors of the insulin-like growth factor-I receptor pathway: lessons from the first clinical trials. Mol Cancer Ther 7:2575–2588

Schuster SJ, Venugopal P, Kern JC et al (2008) GM-CSF plus rituximab immunotherapy: translation of biologic mechanisms into therapy for indolent B-cell lymphomas. Leuk Lymphoma 49:1681–1692

Senter PD (2009) Potent antibody drug conjugates for cancer therapy. Curr Opin Chem Biol 13:235–244

Shen J, Zhu Z (2008) Catumaxomab, a rat/murine hybrid trifunctional bispecific monoclonal antibody for the treatment of cancer. Curr Opin Mol Ther 10:273–284

Shida D, Takabe K, Kapitonov D et al (2008) Targeting SphK1 as a new strategy against cancer. Curr Drug Targets 9:662–673

Sica A, Allavena P, Mantovani A (2008) Cancer related inflammation: the macrophage connection. Cancer Lett 267:204–215

Smyth MJ, Godfrey DI, Trapani JA (2001) A fresh look at tumor immunosurveillance and immunotherapy. Nat Immunol 2:293–299

Takeda K, Stagg J, Yagita H et al (2007) Targeting death-inducing receptors in cancer therapy. Oncogene 26:3745–3757

Tang B, Vu M, Booker T et al (2003) TGF-beta switches from tumor suppressor to prometastatic factor in a model of breast cancer progression. J Clin Invest 112:1116–1124

Toschi L, Janne PA (2008) Single-agent and combination therapeutic strategies to inhibit hepatocyte growth factor/MET signaling in cancer. Clin Cancer Res 14:5941–5946

Tucker GC (2006) Integrins: molecular targets in cancer therapy. Curr Oncol Rep 8:96–103

Wagner K, Schulz P, Scholz A et al (2008) The targeted immunocytokine L19-IL2 efficiently inhibits the growth of orthotopic pancreatic cancer. Clin Cancer Res 14:4951–4960

Walensky LD, Kung AL, Escher I et al (2004) Activation of apoptosis in vivo by a hydrocarbon-stapled BH3 helix. Science 305:1466–1470

Waller EK (2007) The role of sargramostim (rhGM-CSF) as immunotherapy. Oncologist 2:22–26

Wang S (2008) The promise of cancer therapeutics targeting the TNF-related apoptosis-inducing
    ligand and TRAIL receptor pathway. Oncogene 27:6207–6215
Weber J (2008) Overcoming immunologic tolerance to melanoma: targeting CTLA-4 with ipili-
    mumab (MDX-010). Oncologist 4:16–25
Weiss JM, Subleski JJ, Wigginton JM et al (2007) Immunotherapy of cancer by IL-12-based
    cytokine combinations. Expert Opin Biol Ther 7:1705–1721
Wu AM, Senter PD (2005) Arming antibodies: prospects and challenges for immunoconjugates.
    Nat Biotechnol 23:1137–1146
Yang J, Weinberg RA (2008) Epithelial-mesenchymal transition: at the crossroads of development
    and tumor metastasis. Dev Cell 14:818–829

Index

D.A. Frank (ed.), *Signaling Pathways in Cancer Pathogenesis and Therapy*,
DOI 10.1007/978-1-4614-1216-8, © Springer Science+Business Media, LLC 2012